AI. in 2020

A year writing about Artificial Intelligence

Jair Ribeiro

Dedication

To my wife, Gosia, who is my main source of inspiration and serenity in this life.

Epigraph

"Your time is limited, so don't waste it living someone else's life.

Do not be trapped by dogma - which is living with the results of other people's thinking.

Do not let the noise of other's opinions drown out your inner voice.

And most important, have the courage to follow your heart and intuition.

They somehow already know what you truly want to become.

Everything else is secondary."

Steve Jobs

Acknowledgments

Writing a book, large or small, technical or poetic, humble or epic, is always a job that depends on inspiration.
I must thank some people who have contributed to inspire me to research, to develop skills, and to share the most relevant knowledge I have acquired during my life:

First of all, thanks to Malgorzata, my awesome wife, for providing me the daily serenity and energy to work, write, learn, and love.

Thanks to David Bombelli for our 10-years old friendship and the constant professional inspiration and mentorship.
Thanks to Luca Ridolfi, who inspired me with his book: Scattered Notes about Machine Learning, thanks to Mariana Verzaro for our 15-years old friendship that shows me every day the value of resiliency and neuroplasticity.

Thanks to Danilo McGarry, who has inspired me to be a better professional in anything I do since the first day we met, and thanks to Reef Larsson for inspiring me with a passion for sharing knowledge.
And last but not least, thanks to Daniele Zebini, who inspires me with a passion for telling stories and changing the world.

Preface

This book collects the best articles about several artificial intelligence concepts that I have published online during 2020. It is dedicated to anyone interested in Artificial Intelligence and anyone who wants to understand some of the building blocks that form this fascinating technology.

Here, you will find my best articles, updated and revisited, with some more insights, with a suitable format for book readers. The content of this book results from extensive research, long nights of studies, and some of my best years of work in the field in some prestigious enterprise companies in Europe.

My goal is to share as much as possible through an affordable, simple, and straightforward language, valuable knowledge that helps you understanding complex topics related to technologies such as Machine Learning, Deep Learning, Analytics, and Autonomous Vehicles, among others.

It is a satisfying adventure, I must say. Every day I receive considerably positive feedback, lots of article views, lots of likes, retweets, and more on my social networks and not less, some indications as a top writer, invitations to collaborate in some prestigious online publications. All this is truly motivating.

I believe that life is complicated enough, so I consider that every time someone tries to simplify concepts and knowledge useful to humanity, this can be regarded as an essential contribution to inclusiveness and equity in the world. So, this is my mission.

This book is not intended to exhaust all the learning needs of those wishing to enter the AI world. It is a starting point composed of some "scattered notes" that will help you put together some valuable pieces of technology's great mosaic.

The articles presented here are very beneficial to provide you a practical introduction to some of the most important concepts that many of us face daily.

They also will give you some pointers on how to go beyond the first step in search of much more.

Because just as Dante suggested:

"You were not meant to live as ugly, but to seek virtue and knowledge."

Figure 1-a- Allegorical Portrait of Dante - Agnolo Bronzino - Public Domain

2 - People vector created by pch.vector - www.freepik.com

Maybe one day, a robot will steal your job… but there is something you can do today to avoid that.

As for any industrial revolution in the past, millions of people will lose their jobs in the next few years due to Automation. Up to a third of the current work can be automated within the next ten years. This is not a science fiction prediction. It's already happening. Here and now.

Robots are doing today the work of lawyers in a fraction of the time. Journalists are being replaced by software that writes financial reports almost simultaneously when the data is released. Chatbots can compose a winning stock portfolio for you in seconds. Even writers have been replaced by these ever-smarter machines able to learn how to produce the next best-sellers.

Technology enables creating an endless list of innovative products and services that can eliminate entire professions, replace functions, and change work processes.

Robots, self-driving cars, stock-controlling refrigerators, virtual assistants, such as Apple's Siri, are only now possible because artificial intelligence is evolving on a speedy scale.

It is increasingly evident that all professionals will need to adapt to a scenario with new technologies, robots, and Artificial Intelligence in several areas as their occupations suffer from this revolution. Some people will be able to surf these waves through better education. Others will dedicate their time and energy to activities that require emotional and social skills, creativity, a high level of cognitive ability, and difficult skills for robots to replicate.

Those who will suffer most of the massive Automation effects are the professionals who occupy machine operators and food industry employees. Nor are they immune to Automation, real estate brokers, legal assistants, accountants, and professionals from administrative sectors.

On the other hand, jobs that require human interaction, such as doctors, lawyers, teachers, and bartenders, are less likely to be replaced by robots. Skilled jobs, but not very high salaries, such as gardeners, plumbers, and caregivers, are also less vulnerable. And several new activities, new jobs, new professions, and new skills will be created by the evolution of robots and Artificial Intelligence.

In the next few years, the world will experience a transition on the scale of the one that occurred in the early 1900s. Industrial development transformed much of the work, primarily agricultural, and new jobs and opportunities created by technology enhancement.

But to survive to this A.I. is every day more clear that we need to develop our interest in arts and sciences in our professional education, benefiting our I.Q. and our ability for more complex social relationships because this is something machines cannot do better than humans … well … at least for a long time.

Statistical studies carried out by the European Journal of Personality, with more than 340,000 people over 50, found every

15 points of I.Q. Found, it is possible to reduce by 7% the risk of a professional being replaced by a machine.

Applying this conclusion to the United States' entire population, we can estimate that more than 10 million jobs can be saved amongst people threatened by Automation. In the specific case of scientific activities, the study found that, on a 5-point scale, a 1-point increase in taste for science-related jobs would guarantee more than 3 million jobs threatened by machines.

The leaders of the future

Not only the professions, the advent of Artificial Intelligence and Automation also changes the concept of leadership.

The leaders of this automated future need to see the size of the transformation. They must know that technology will be increasingly required and used to empower people, ensuring that it is impossible to take the first step by eliminating people and including them.

To embrace this new reality, we need a radical readaptation of the entire education system to emphasize personality traits and promote more significant social interaction in our schools, developing more and more skills that machines can not emulate.

Humans will always overcome the machines when it comes to tasks that require creativity and a degree of complexity that runs out of routine, requiring flexibility, ambiguity, and improvisation.

In this future of intelligent machines, our socioeconomic origins will not define our professional perspectives, which will be determined by our intelligence levels, extraversion, maturity, and the degree of interest in arts and sciences.

So, what do you can do to keep your work in the future?

Suppose you are worried about how to face your professional future during the fourth industrial revolution. In that case, I see two possible paths for you:

The first is to be an entrepreneur. Do what you love, do what you want, and do what makes sense to you in your life, turning it into a business, preferably creating something that machines cannot do better than humans.

The second alternative is to decide to take a turn in your career, invest your time and energy in one of the new professional areas created by A.I. It is not about throwing away the experience gained over the years. Still, it is about changing your mindset, tacit knowledge, networking, and eventually requalifying yourself as a professional. It's never too early or never too late for a career turnaround.

And whatever road you decide to take, Artificial Intelligence will be there, so we all need to update and empower ourselves to handle related technologies, but it is essential to have a clear purpose. The times are changing. Suppose yesterday the focus was on creativity, leadership, problem-solving skills, adaptation to context, and, of course, Emotional Intelligence (it is fundamental always), today, with the strength of Automation and Artificial Intelligence. In that case, there is a reversal, and the skills are other in this scenario.

In the next future, collaboration, perspective, and purpose will the most valuable skills to demonstrate how good a human being is to transform reality better than any machine.

By Jair Ribeiro on January 2, 2020.

3- People vector created by freepik - www.freepik.com

Stop everything you are doing and watch these five TED Talks on A.I. Ethics now.

Some of the most impactful "food for thoughts" TED Talks about A.I. Ethics can make you start thinking about it today.

A.I. has a significant impact on our daily lives. It should require us a better understanding of the positive and negative effects of this technology.

The ethical challenges that artificial intelligence poses in our lives today are becoming well known. It's time to understand better how this technology's ethical aspects can be systematized in a realistic and enforceable way.

Today A.I. impacts our jobs, safety, shopping, justice, and several other activities. In many cases, all of this is happening without a shared and well defined ethical and legal structure to ensure that the technology below it is transparent, accountable, and responsible.

We are now in 2020, and I firmly believe that this should start the change. We should begin to pay attention and consider it an emergency at the same climate change level today.

I've been dedicating a consistent part of my studies, my public speaking in conferences, and my networking to promoting awareness and call-to-action regarding the necessity to bring up A.I. Ethics on every possible occasion when technology frameworks have been discussed.

And from today, you will find on my Medium more and more articles about A.I. Ethics in a practical and impactful way.

To start, I would like to share some of my favorite "food for thoughts" TED Talks about A.I. Ethics that can make you start thinking about it.

I've selected five thought leaders who share what they consider necessary to govern A.I. and ensure a comprehensive ethical framework, in addition to human intelligence.

How to Keep Human Bias Out of AI

AI algorithms make important decisions about you all the time—like how much you should pay for car insurance or whether or not you get that job interview.

But what happens when these machines are built with human bias coded into their systems? Technologist Kriti Sharma explores how the lack of diversity in tech is creeping into our A.I., offering three ways we can start making more ethical algorithms.

Kriti Sharma is the Founder of A.I. for Good, an organization focused on building scalable technology solutions for social good. In 2018, she also launched rAInbow, a digital companion for women facing domestic violence in South Africa. This service reached nearly 200,000 conversations within the first 100 days, breaking down gender-based violence stigma. In 2019, she collaborated with India's Population Foundation to launch Dr.

Sneha, an AI-powered digital character to engage young people in sexual health. This issue is still considered taboo in India.

Sharma was recently named in the *Forbes* "30 Under 30" list for advancements in A.I. She was appointed a United Nations Young Leader in 2018 and is an advisor to both the United Nations Technology Innovation Labs and the U.K. Government's Centre for Data Ethics and Innovation.

Can We Protect A.I. from Our Biases?

As humans, we're inherently biased. Sometimes it's explicit. Other times it's unconscious, but as we move forward with technology, how do we keep our biases out of the algorithms we create? Documentary filmmaker Robin Hauser argues that we need to have a conversation about how A.I. should be governed and ask who is responsible for overseeing these supercomputers' ethical standards. "We need to figure this out now," she says. "Because once skewed data gets into deep learning machines, it's challenging to take it out."

Robin is the director and producer of cause-based documentary films at Finish Line Features, Inc. and Unleashed Productions, Inc. As a businesswoman, long-time professional photographer, and social entrepreneur, Robin brings her leadership skills, creative eye, and passion for her documentary film projects. Her artistic vision and experience in the business world afford her a unique perspective on what it takes to motivate an audience.

Her most recent award-winning film, CODE: Debugging the Gender Gap, premiered at Tribeca Film Festival 2015 and has caught the eye of the international tech industry and policymakers and educators in Washington, DC, and abroad. Robin is currently directing and producing bias, a documentary about unconscious bias and how it affects our lives socially and in the workplace.

The Era of Blind Faith in Big Data Must End

Algorithms decide who gets a loan, gets a job interview, gets insurance, and much more—but they don't automatically make things fair. Mathematician and data scientist Cathy O'Neil coined a term for algorithms that are secret, important, and harmful: "weapons of math destruction." Learn more about the hidden agendas behind the formulas.

In 2008, as a hedge-fund quant, mathematician Cathy O'Neil saw firsthand how bad math could lead to financial disaster. Disillusioned, O'Neil became a data scientist and eventually joined Occupy Wall Street's Alternative Banking Group.

With her popular blog mathbabe.org, O'Neil emerged as an investigative journalist. Her acclaimed book *Weapons of Math Destruction* details how opaque, black-box algorithms rely on biased historical data to do everything from sentence defendants to hire workers. In 2017, O'Neil founded consulting firm ORCAA to audit algorithms for racial, gender, and economic inequality.

Machine Intelligence Makes Human Morals More Important

Machine intelligence is here, and we're already using it to make subjective decisions. But the complex way A.I. grows and improves makes it hard to understand and even harder to control. In this cautionary talk, techno-sociologist Zeynep Tufekci explains how intelligent machines can fail in ways that don't fit human error patterns—and in ways, we won't expect or be prepared for. "We cannot outsource our responsibilities to machines," she says. "We must hold on ever tighter to human values and human ethics."

We've entered an era of digital connectivity and machine intelligence. Complex algorithms are increasingly used to make consequential decisions about us. Many of these decisions are subjective and have no right answer: who should be hired, fired, or promoted, what news should be shown to whom your friends do you see updates from, should be paroled. With the increasing use of machine learning in these systems, we often don't even understand how they make these decisions. Zeynep Tufekci studies what this historic transition means for culture, markets, politics, and personal life.

Tufekci is a contributing opinion writer at the *New York Times*, an associate professor at the School of Information and Library Science at the University of North Carolina, Chapel Hill, and a faculty associate at Harvard's Berkman Klein Center for Internet and Society.

Her book, *Twitter and Tear Gas: The Power and Fragility of Networked Protest*, was published in 2017 by Yale University Press. From Penguin Random House, her next book will be about algorithms that watch, judge, and nudge us.

How to Get Empowered Not Overpowered

Many artificial intelligence researchers expect A.I. to outsmart humans at all tasks and jobs within decades, enabling a future where we're restricted only by the laws of physics, not the limits of our intelligence. MIT physicist and A.I. researcher Max Tegmark separate the real opportunities and threats from the myths, describing the concrete steps we should take today to ensure that A.I. ends up being the best—rather than the worst—thing to ever happen to humanity.

Max Tegmark is an MIT professor who loves thinking about life's big questions. He's written two popular books, *Our Mathematical Universe: My Quest for the Ultimate Nature of Reality* and the recently published *Life 3.0: Being Human in the Age of Artificial Intelligence,* as well as more than 200 nerdy technical papers on topics from cosmology to A.I.

It's time for action

I know that we have some pretty complicated questions to answer regarding A.I.'s ethical framework. Often, there are no simple answers. After all, by definition, ethical dilemmas do not have clearly defined answers, just like everything related to morality;

that's why we must discuss them to agree on a consensus with the right urgency and awareness.

A.I. is transforming our society, and it can't just be the privilege of a few to decide how that will happen or frame what that world looks like.

Transparency is fundamental when there is bias, even though it is unintended. The technology leaders' responsibility is to open up the black box of A.I. and ensure there is as much transparency as possible.

And we have our part of doing. Let's start from here?

By Jair Ribeiro on January 13, 2020.

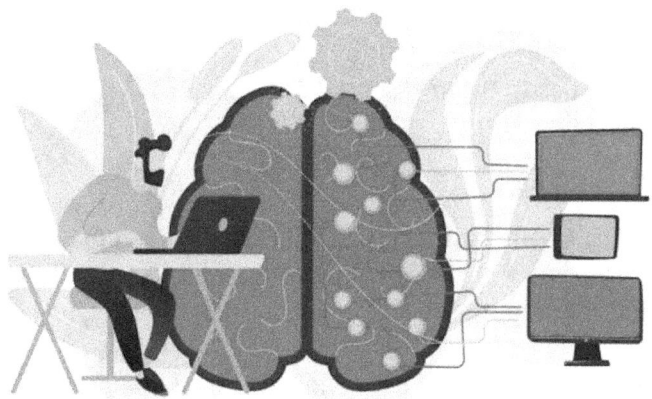

4 - Education vector created by vectorjuice - www.freepik.com

An easy guide to the history of Artificial Intelligence

A quick look at some of the most critical A.I. events since its beginning and some interesting links.

Suppose we start to coming back in history... until ancient Greek. In that case, we can discover that intelligent machines and artificial beings first appeared as myths of Antiquity.

But when it comes to A.I. and Machine Learning, we don't go so far with the memory because the history of artificial intelligence as we think of it today spans less than a century. I want to share a quick look at some of the most critical A.I. events since its beginning and some interesting links.

1943

Warren McCullough and Walter Pitts published "A Logical Calculus of Ideas Immanent in Nervous Activity." The paper proposed the first mathematic model for building a neural network.

1949

In his book The Organization of Behavior: A Neuropsychological Theory, Donald Hebb offers the theory that neural pathways are created from experiences and that connections between neurons become stronger the more frequently they're used. Hebbian learning continues to be an essential model in A.I.

1950

Alan Turing publishes "Computing Machinery and Intelligence, proposing what is now known as the Turing Test, a method for determining if a machine is intelligent.

Harvard undergraduates Marvin Minsky and Dean Edmonds build SNARC, the first neural network computer.

Claude Shannon publishes the paper "Programming a Computer for Playing Chess."

Isaac Asimov publishes the "Three Laws of Robotics."

1952

Arthur Samuel develops a self-learning program to play checkers.

1954

The Georgetown-IBM machine translation experiment automatically translates 60 carefully selected Russian sentences into English.

1956

The phrase artificial intelligence is coined at the "Dartmouth Summer Research Project on Artificial Intelligence." Led by John McCarthy, the conference, which defined A.I.'s scope and goals, is widely considered to be the birth of artificial intelligence as we know it today.

Allen Newell and Herbert Simon demonstrate Logic Theorist (LT), the first reasoning program.

1958

John McCarthy develops the A.I. programming language Lisp and publishes the paper "Programs with Common Sense." The paper proposed the hypothetical Advice Taker, a complete A.I. system with the ability to learn from experience effectively as humans do.

1959

Allen Newell, Herbert Simon, and J.C. Shaw develop the General Problem Solver (GPS), a program designed to imitate human problem-solving.

Herbert Gelernter develops the Geometry Theorem Prover program.

Arthur Samuel coins the term machine learning while at IBM.

John McCarthy and Marvin Minsky found the MIT Artificial Intelligence Project.

1963

John McCarthy starts the A.I. Lab at Stanford.

1966

The Automatic Language Processing Advisory Committee (ALPAC) report by the U.S. government details the lack of progress in machine translation research, a major Cold War initiative with the promise of automatic and instantaneous Russian translation. The ALPAC report leads to the cancellation of all government-funded M.T. projects.

1969

The first successful expert systems are developed in DENDRAL, a XX program, and MYCIN, designed to diagnose blood infections, are created at Stanford.

1972

The logic programming language PROLOG is created.

1973

The "Lighthill Report," detailing the disappointments in A.I. research, is released by the British government and leads to severe funding cuts for artificial intelligence projects.

1974–1980

Frustration with the progress of A.I. development leads to major DARPA cutbacks in academic grants. Combined with the earlier ALPAC report and the previous year's "Lighthill Report," artificial intelligence funding dries up and research stalls. This period is known as the "First AI Winter."

1980

Digital Equipment Corporations develop R1 (also known as XCON), the first successful commercial expert system. Designed to configure orders for new computer systems, R1 kicks off an investment boom in expert systems that will last for much of the decade, effectively ending the first "AI Winter."

1982

Japan's Ministry of International Trade and Industry launches the ambitious Fifth Generation Computer Systems project. The goal of FGCS is to develop supercomputer-like performance and a platform for A.I. development.

1983

In response to Japan's FGCS, the U.S. government launches the Strategic Computing Initiative to provide DARPA funded research in advanced computing and artificial intelligence.

1985

Companies are spending more than a billion dollars a year on expert systems, and an entire industry known as the Lisp machine market springs up to support them. Companies like Symbolics and Lisp Machines Inc. build specialized computers to run on the A.I. programming language Lisp.

1987–1993

As computing technology improved, cheaper alternatives emerged, and the Lisp machine market collapsed in 1987, ushering in the "Second AI Winter." During this period, expert systems proved too expensive to maintain and update, eventually falling out of favor.

Japan terminates the FGCS project in 1992, citing failure in meeting the ambitious goals outlined a decade earlier.

DARPA ends the Strategic Computing Initiative in 1993 after spending nearly $1 billion and falling far beyond expectations.

1991

U.S. forces deploy DART, an automated logistics planning and scheduling tool, during the Gulf War.

1997

IBM's Deep Blue beats world chess champion, Gary Kasparov

2005

STANLEY, a self-driving car, wins the DARPA Grand Challenge.

The U.S. military begins investing in autonomous robots like Boston Dynamic's "Big Dog" and iRobot's "PackBot."

2008

Google makes breakthroughs in speech recognition and introduces the feature in its iPhone app.

2011

IBM's Watson trounces the competition on Jeopardy!

2012

Andrew Ng, the founder of the Google Brain Deep Learning project, feeds a neural network using deep learning algorithms 10 million YouTube videos as a training set. The neural network learned to recognize a cat without being told what a cat is, ushering in a breakthrough era for neural networks and deep learning funding.

2014

Google makes the first self-driving car to pass a state driving test.

2016

Google DeepMind's AlphaGo defeats world champion Go player Lee Sedol. The complexity of the ancient Chinese game was seen as a significant hurdle to clear in A.I.

2017

In October 2017, Sophia, a social humanoid robot developed by Hong-Kong based company Hanson Robotics, became the first robot to receive citizenship of any country and named the United Nations Development Programme's first-ever Innovation Champion and is the first non-human to be given any United Nations title.

2018

Jair Ribeiro leaves IBM to become a Senior A.I. Business Analyst at the A.I. & ML Center of Excellence in Volvo Group—maybe one day it will be written in history books. :-)

2019—Yoshua Bengio, Geoffrey Hinton, and Yann LeCun, the godfather of modern A.I., won the Turin Award for their work developing the A.I. subfield deep learning.

2020

What breakthrough in 2020 do you think will enter this list?

By Jair Ribeiro on January 16, 2020.

5 - Abstract vector created by pch.vector - www.freepik.com

Thirty-eight free courses to help you master the most in-demand job skills in 2020.

What skills does the workforce value most in 2020, and how can you learn them today? For free!

Every year, LinkedIn analyzes data from its network of over 660+ million professionals and 20+ million jobs to reveal the 15 most in-demand soft and hard skills of this year and release *free courses to help you learn them on LinkedIn Learning.*

This time, Blockchain has topped the list of skills companies are looking for in employees around the world this year.

Blockchain was born in 2009 to support the use of cryptocurrency. But Blockchain's novel way to store, validate, authorize, and move data across the internet has evolved to store and send any digital asset securely. The small supply of professionals who have this skill is in high demand.

Blockchain was the top priority for employers hiring in the U.S., U.K., France, Germany, and Australia. Yet, both the first-time Blockchain made it onto LinkedIn's rankings of in-demand skills and came first.

Industries outside the financial services sector were increasingly seeking talent with Blockchain experience, including retail, shipping, healthcare, farming, and gaming.

LinkedIn measured demand by looking at its users' profiles to determine the frequency that people with different skillsets were getting hired.

Cloud computing came in second place, which allows data to be stored and managed on the internet.

Artificial intelligence (A.I.), the technology developing machine-learning, was the fourth most in-demand area of "hard" skills.

Artificial intelligence (A.I.) augments the capabilities of the human workforce. The people who can harness the power of A.I., machine learning, and natural language processing are the ones who will help organizations deliver more relevant, personalized, and innovative products and services.

Rounding out the top five was UX design, focusing on users' experience of products, particularly technology.

When it comes to UX, it seems like consumers' average attention span decreases every year. They have little patience for products that aren't intuitive.

Organizations need more expertise to help them build more human-centric products and experiences.

Here you have the top 10 most in-demand hard skills globally and some very interesting training courses you can have on LinkedIn Learning to help you to develop these skills:

1. Blockchain
2. Cloud computing
3. Analytical reasoning
4. Artificial intelligence
5. UX design
6. Business analysis

7. Affiliate marketing
8. Sales
9. Scientific computing
10. Video production

#1 Blockchain—New

Blockchain was born in 2009 to support the use of cryptocurrency. But Blockchain's novel way to store, validate, authorize, and move data across the internet has evolved to store and send any digital asset securely. The small supply of professionals who have this skill is in high demand.

Learn Blockchain in this course—free through January 31:

- Blockchain Basics with Jonathan Reichental

More recommended courses:

- Blockchain Beyond the Basics with Jonathan Reichental
- Blockchain: Learning Solidity with Emmanual Henri

#2 Cloud Computing—Down 1

Today, companies are built and run on the cloud. They need the talent to help them drive technical architecture, design, and delivery of cloud systems like Microsoft Azure.

Learn cloud computing in this course—free through January 31:

- Learn Cloud Computing: Core Concepts with David Linthicum

More recommended courses:

- Azure Administration Essential Training with David Elfassy
- Cloud-Native Development with Chris Bailey

#3 Analytical Reasoning—Same as 2019

Data has become the foundation of every single business. Organizations want talent to make sense of it and uncover insights that drive the company's best decisions.

Learn analytical reasoning in this course—free through January 31:

- Strategic Thinking with Dorie Clark

More recommended courses:

- Learning Data Analytics with Robin Hunt
- Power B.I. Top Skills with John David Ariansen and Madecraft

#4 Artificial Intelligence—Down 2

Artificial intelligence (A.I.) augments the capabilities of the human workforce. The people who can harness the power of A.I., machine learning, and natural language processing are the ones who will help organizations deliver more relevant, personalized, and innovative products and services.

Learn Artificial Intelligence in this course—free through January 31:

- Artificial Intelligence Foundations: Machine Learning with Doug Rose

More recommended courses:

- Big Data in the Age of A.I. with Barton Poulson
- Introducing A.I. to Your Organization with Jonathan Fernandes

#5 UX Design—Same as 2019

It seems like consumers' average attention span decreases every year, and they have little patience for products that aren't intuitive. Organizations need more expertise to help them build more human-centric products and experiences.

Learn UX design this course—free through January 31:

- Getting Started in User Experience with Chris Nodder

More recommended courses:

- Learning Adobe X.D. with Tom Green
- Interaction Design: Software and Web Design Patterns with Diane Cronenwett

#6 Business Analysis—Up 10

The business analysis made the most significant jump of any skill on our list. It's one of the few hard skills every professional should have, as most roles require some business analysis to make decisions.

Learn business analysis in this course—free through January 31:

- Business Analysis Foundations with Greta Blash

More recommended courses:

- Data Analytics for Business Professionals with John Johnson
- Data-Driven Presentations with Excel and PowerPoint with Gigi von Courtner

#7 Affiliate Marketing—New

With the decline of traditional advertising and social media, affiliate marketing is rapidly rising as a must-have hard skill. Affiliate marketing leverages company partnerships or influencers that are hyper-targeted to a particular audience.

Learn affiliate marketing in this course—free through January 31:

- Influencer Marketing Foundations with Chelsea Krost

More recommended courses:

- Marketing Tools: Digital Marketing with Anson Alexander
- Improve SEO for your Ecommerce Site with Sam Dey

#8 Sales—Same as 2019

You'd be hard-pressed to find a company that doesn't need great salespeople—those who can effectively manage a sales team, understand the sales funnel, work with cross-functional partners, and sell into the highest levels of the business.

Learn sales in this course—free through January 31:

- Social Selling Foundations with Derek Pando

More recommended Courses:

- Cross-Functional Sales Teams with Jeff Bloomfield
- Sales Enablement with Meridith Powell

#9 Scientific Computing—Up 3

Scientific computing skills are held by data science professionals, engineers, and software architects, and others. Companies need more professionals who can develop machine learning models and apply statistical and analytical approaches to large data sets using Python, MATLAB, and others.

Learn scientific computing in this course—free through January 31:

- Parallel and Concurrent Programming with Python 1 with Barron Stone and Olivia Chiu Stone

More recommended courses:

- Learning MATLAB with Steven Moser
- Introduction to Quantum Computing with Jonathan Reichental

#10 Video Production—Down 3

Consumers have an insatiable appetite for video content, making sense that video production continues to be a priority for companies. Cisco estimates that video will account for 82% of global internet traffic in 2022.

Learn video production in this course—free through January 31:

- Social Media Video Strategy: Weekly Bites with Ashley Kennedy

More recommended courses:

- Connecting with Your Audience Using Video with Jaime Cohen
- Social Media Video for Business and Marketing with Ashley Kennedy

Not only Hard skills...

LinkedIn also ranked "soft" skills—the essential interpersonal skills that make or break our ability to get things done in our current jobs and take on new opportunities ahead.

Topping this year's list are creativity, collaboration, persuasion, and emotional intelligence—all skills that demonstrate how we work with others and bring new ideas to the table.

Four of the five most in-demand soft skills remain in their top spots year over year, further reinforcing that these skills are evergreen—they're likely to stay the top skills that companies want in star employees.

The list looked very similar to the 2019 rankings, with creativity holding onto the top spot. The one variation in the most in-demand soft skills list indicates that companies are gravitating toward talent with interpersonal and people-oriented skills.

'Time management,' a more task-oriented skill, fell off the top soft skills list. 'Emotional intelligence,' the ability to perceive, evaluate, and respond to both your own emotions and those of others, took its place. This emphasized the "importance of how we react to and interact with colleagues.

While task-oriented skills remain critical to our success at work, the data shows that employers value our ability to work well with colleagues.

Here you have the Top 5 most in-demand soft skills globally and some exciting training courses you can follow in January for free to develop the most demanded skills for 2020:

1. Creativity
2. Persuasion
3. Collaboration
4. Adaptability
5. Emotional intelligence

For the entire month of January, LinkedIn Learning unlocked courses that will help you hone these highly sought after skills—for free.

#1 Creativity—Same as 2019

Organizations need people who can creatively approach problems and tasks across all business roles, from software engineering to H.R. Focus on honing your ability to bring new ideas to the table in 2020.

Learn creativity in this course—free through January 31:

- Banish Your Inner Critic to Unleash Creativity with Denise Jacobs

More recommended courses:

- Creativity For All (Weekly Series)
- Creative Exercises to Spark Original Thinking with Amy Wynne

#2 Persuasion—Same as 2019

Leaders and hiring managers value individuals who can explain the "why." To advance your career, brush up on your ability to effectively communicate ideas and persuade your colleagues and stakeholders that it's in their best interest to follow your lead.

Learn persuasion in this course—free through January 31:

- Persuading Others with Dorie Clark

More recommended courses:

- Leading Without Formal Authority with Elizabeth (McLeod) Lotardo and Lisa Earle McLeod
- Persuasive Coaching with Brian Ahearn

#3 Collaboration—Same as 2019

High-functioning teams can accomplish more than any individual—and organizations know it. Learn how your strengths can complement those of your colleagues to reach a common goal.

Learn collaboration in this course—free through January 31:

- Being an Effective Team Member with Daisy Lovelace

More recommended courses:

- Shane Snow on Dream Teams
- Teamwork Foundations with Chris Croft

#4 Adaptability—Same as 2019

The only constant in life—and business—is change. To stand out in 2020, embrace that reality and make sure to show up with a positive attitude and open-minded professionalism, especially in stressful situations.

Learn adaptability in this course—free through January 31:

- Managing Stress for Positive Change with Heidi Hanna

More recommended courses:

- Developing Adaptability as a Manager with Dorie Clark
- Finding Your Time Management Style with Dave Crenshaw

#5 Emotional Intelligence—New

Emotional intelligence is the ability to perceive, evaluate, and respond to your own emotions and the emotions of others. New to the most in-demand skills list this year, the need for emotional intelligence underscores the importance of effectively responding to and interacting with our colleagues.

Learn emotional intelligence in this course—free through January 31:

- Developing Your Emotional Intelligence with Gemma Leigh Roberts

More recommended courses:

- Social Success at Work with Todd Dewett
- Influencing Others with John Ullman

Now that you made your deepdive into the 2020 list, you can start learning the skills companies need most. I hope that with insight into what companies need today, you feel ready to cultivate the essential soft skills and hard skills and empower to own your career.

By Jair Ribeiro on January 20, 2020.

6 - People vector created by pch.vector - www.freepik.com

Can "fake faces" Lead to the Illusion of Diversity?

GANs—or Generative Adversarial Networks have been used to create agency photos with non-existent faces that promote diversity, generating business opportunities. However, this use of A.I. can also have numerous ethical implications.

Artificial intelligence is presenting considerable advancements in the generation of compelling and fascinating images of unquestionably realistic features that are almost impossible to be classified as belonging to people who do not exist anywhere in the world. If we compare the faces produced no more than five years ago and those published recently, the improvements are incredible.

With the new GANs—Generative Adversarial Networks, algorithmic architectures that use two neural networks, competing one against the other (thus the term "adversarial") to generate new, synthetic instances of data, these synthetic faces are easily customizable and editable, making them so credible thanks to particular effects.

Applying a mix of features from real images such as the shades of the skin and the hair color, for example, to the fake ones is possible to generate a virtual population that does not exist in the real world.

A business opportunity

Why use images of non-existent people, created by an algorithm, instead of photos of real people, can be so exciting?

One of the reasons is first and foremost economic: once the algorithm training is completed, carried out using large datasets of real photographs, A.I. can stir out immense quantities of images of fake people in every possible situation and with expressions of any kind, inevitably lowering the prices for "real models" and real photographic production.

Now we should start to ask ourselves if we've come this far in the last five years, where will we be five years from now? How many more domains will A.I. rule that today are considered exclusive of humans?

This can be very useful to those who need to increase promotional material quickly, prepare numerous drafts, or illustrate concepts that could be time and money consuming for a human model.

Imagine one agency that needs a glamorous girl with blond hair and green eyes? Among the tens of thousands of images—viewable on the GeneratedPhotos website—there is likely one that is right for you.

However, a second reason for these services' birth is much less intuitive. It has everything to do with the historical lack of diversity of advertising images and the like. The prevalence of white men and women is overwhelming. This is a problem that also can be at the origin of some forms of algorithmic bias).

In general, in the databases of photo agencies, there is an evident prevalence of white men. Simultaneously, minorities are frequently under-represented, putting those looking for images characterized by a precise diversity in some difficulty in finding them. And it is here that, once again, technologies like the GANs, which can quickly produce thousands of fake photographs with a very high percentage of diversity, can be applied.

Diversity is a very delicate topic, and I wonder if artificial intelligence is ready to solve it. We risk building a false illusion of diversity, considering that artificial intelligence can inherit conscious or unconscious bias presented to it during the algorithm's training, resulting in increasing the homogeneity instead of increasing the diversity in the real world.

From the ethical point of view, this technology's use for this specific purpose should raise several discussions about the truthfulness of what we see online today and in the future.

This ethical consideration is valid for "fake images" in the same way it is valid for the "deepfakes" or "fake news" since they use the same technology called GAN, a generative adversarial network.

Artificial images and videos created by GANs are called deepfakes. They've been widely discussed in the news, primarily when used maliciously.

As we saw in other areas where A.I. made quite impressive advances during the last years, we risk with the "fake images" to finding ourselves in a world in which distinguishing reality from fiction becomes increasingly tricky.

Maybe we are very close to discover that artificial intelligence is also managing to automate a job that until yesterday seemed to be reserved only for humans: the model.

Inverting the roles: Not only A.I. faces but also A.I. beauty contests

If this is not worrying enough, and if it seems unconnected somehow, some years, the far-away 2016, we had the first international beauty contest judged by "machines." The software is supposed to use objective factors such as facial symmetry and wrinkles in identifying the most attractive contestants.

The contest, named Beauty A.I., had the participation of thousands of people from several countries who submitted their photographs in the expectation that artificial intelligence, supported by complex algorithms, would determine that their faces better resembled the classic "human beauty."

But once the results came out, the creators were surprised to see that there was a glaring factor linking the winners: the robots did not like people with dark skin.

Out of 44 winners, nearly all were white, a handful was Asian, and only one had dark skin. That's even though, although the majority of contestants were white, many people of color submitted photos, including large groups from India and Africa.

The ensuing controversy has sparked renewed debates about how algorithms can perpetuate biases, yielding unintended and often offensive results.

There may be thousands of reasons for this kind of algorithm to behave like that, preferring light skin. Still, it's a technical fact that these algorithms often rely on large datasets of photos to be trained. The data used to establish standards of attractiveness did not include enough diversity. It's simple: If you do not have enough diversity within the dataset, you might have biased results.

> Often the simplest explanation for biased algorithms is that the humans who create them have their own deeply entrenched biases. Despite perceptions that algorithms are somehow neutral and uniquely objective, they can often reproduce and amplify existing prejudices.

I consider GANs one of the most powerful new machine learning technologies available today. Still, we must be aware of the ethical implications since techniques like these are already being used, for example, by people impersonating journalists on Twitter with AI-generated profile pictures and online tools for generating fake profiles. But can GANs be used for good?

Is it possible to use GANs for good today?

There are many opportunities to apply GANs to create useful, powerful, and ethical tools. For instance, researchers have created a project where they have programmed GANs to perform encryption and decryption forms and apply these operations selectively to meet confidentiality goals with their cryptography systems.

But not only, but GANs can also be used in several other interesting ways as you can see:

Fighting bias with GANs

Surprisingly, the same technology used to create fake images can be used to detect bias in algorithms. Adversarial neural networks can be used to ensure that an A.I. system is cleared from racial bias. Imagine one A.I. fed with data regarding people's inherent criminal activities and figuring out what jail sentence they should be given. A second A.I. (Adversarial) could be used to identify potential bias related to race or gender. For example, from the predicted sentence, feedback this data to make the system more balanced and fair.

Solving privacy issues with Synthetic data.

Generative models may be used to synthesize private training data in a way that is indistinguishable from real data. The algorithm can learn how to create synthetic information that maintains real input format and statistical features. However, unlike real records, the generated data does not represent privacy issues since they do not belong to any specif person.

The resulting records allow the training of accurate machine learning models while protecting people's privacy.

Also, GANs can be used to generate 3D objects. They can be used on music generation and be used on tumor detection. Several other applications are under development at this moment.

Conclusion

This kind of technology is still in its infancy. As for any child development issues, the experts pushing the edges of this "baby" have enormous power and responsibility.

As a technology enthusiast and an and ethical A.I. practitioner A.I., I try to keep my eyes open to the latest advances in generative adversarial networks and the opportunities this technology can bring if developed for good.

We must be conscious that we are developing tools and techniques that have the potential to transform our worlds, and it's our responsibility to keep users safe and informed.

By Jair Ribeiro on January 27, 2020.

7 - People vector created by stories - www.freepik.com

Can Artificial Intelligence predict the next pandemic?

While the coronavirus outbreak is growing and thousands of people are getting infected, companies have been using A.I. to predict the next epidemics.

According to WHS's latest figures, the coronavirus outbreak began to take over newspapers in the last week. Today (30/01/2020), the total number of deaths from the virus had risen to 170 in China.

More than 7,700 people have been sickened worldwide, according to Chinese officials and the World Health Organization.

Other than China, at least 14 countries have had confirmed cases, including Japan, Singapore, the United States, France, Germany, Australia, South Korea, and Saudi Arabia.

But could this epidemic outrage be predicted?

Last December, an artificial intelligence system issued a warning about a possible spread of the coronavirus more than a week before the official statement from the World Health Organization (WHO) by a company called BlueDot founded in 2014 that has already raised more than $ 9.4 million in investments.

The algorithm of BlueDot, sent an email to health and airline organizations on December 31, warning them to avoid the area of Wuhan, China, which was later confirmed as the epicenter of the epidemic.

The technology works by mapping news from hundreds of international sources in 65 languages and tracking health research networks, airline data, official announcements from companies related to agribusiness, livestock, and others, and even discussion forums on different topics.

With this huge amount of information, BlueDot consolidates alerts related to possible areas of risk and new diseases, such as the coronavirus, about which the company's customers were informed ten days before the official statement of WHO and seven before the Center for Prevention and United States Disease Control, considered the first to make the epidemic official.

More than just sending warnings about the coronavirus, the system could also predict the path that the infection would follow. Based on data from airlines and flights out of Wuhan, the technology indicated cities like Seoul, Taipei, Tokyo, and Bangkok as outbreaks of the disease before cases were confirmed in all of them. Once again, the alerts aimed to prevent customers from traveling to such regions and sanitary measures to reduce contact with the pathogen.

This kind of warning can be made even earlier if systems like the algorithm developed by BlueDot also consider social networks, which does not happen due to high data pollution, which hinders a reliable analysis. Even so, this usage of A.I. is currently the fastest in detecting this type of epidemic, serving not only to safeguard people's health but also to help contain and treat those infected.

A.I. as a disruptive wave in healthcare

A.I. and Machine Learning systems are undoubtedly the future of healthcare, disrupting the industry for good.

According to Frost & Sullivan, AI systems are projected to be a 6 billion dollar industry by 2021. A recent McKinsey study predicted healthcare as one of the top 5 industries with more than 50 use cases that would involve A.I. and over USD 1bn already raised in startup investments.

A.I. in healthcare will fundamentally impact patients, doctors, administration, and operations.

While A.I. tools and bots have been implemented at every level of a patient's medical journey, it is the overall impact it makes that truly disrupts the industry.

A.I. is seamlessly bringing together all records of a patient, along with insights, use that data for diagnoses, in turn for treatment, and eventually for maintenance of health.

Conclusion

As the world changes rapidly, these diseases are emerging and spreading at a fast pace. However, with A.I. tools and various software in place, the increased access to data can be put to good use.

The substantial increase in data can be used to generate critical insights and, in turn, act on them, thereby spreading news earlier and faster than the disease can spread itself.

The idea of A.I. battling deadly disease offers a case where we might feel slightly less uneasy, if not altogether hopeful. Perhaps this technology—if developed and used correctly—could help save some lives.

By Jair Ribeiro on January 30, 2020.

8 - People vector created by freepik - www.freepik.com

Three intelligent apps that make you more productive today

Here you have a starter kit of Productivity apps that will help you to get more work done in less time.

As one of my previous mentors in IBM used to say, "one thing is to get your job done, but even more important is that you get it done with quality and on time," For me, that used to means: productivity.

I know that there are thousands of apps and tools out there that promise to help you become more productive. Probably there are millions of articles about productivity apps across the google-sphere. Still, I want to share this one here because "by experience," I firmly believe that it will help you get your work done faster and effectively.

Over the last years, I've been developing a very intensive daily routine, trying to balance my personal life as a father-of-three-daughters and a relatively active and multitasking enterprise professional.

And one thing that helps keep my productivity level up is leveraging the right technology to support me in every task.

With this mindset, I've started to use dozens of applications, mostly mobile apps, that help me to get things done faster and with quality, allowing me to have more time in my day for the things I need and love to do.

Also, I have followed different productivity philosophies, always thinking that I should not do more than I have to.

After all, what I've concluded is that one of the best ways to increase my efficiency at work and in my personal life is to cut out things that don't need to be done in the first place.

Therefore, in general, my advice is not to choose apps and tools only because they look fancy but choose those that can help you cut out everything unnecessary, allowing you to focus on solely what you need the most.

To start, I've selected three of the most useful productivity tools in my colossal list to give you a quick address on how to enhance your capacity to get things done quickly and efficiently:

Calendar

Let's start from the basics: if you are looking to save more time, be more productive. If you are trying to achieve better focus, you probably need to start with a smart calendar application.

I know that there are zillions of calendars available on our smartphones today. Most of us are happy with Outlook or Google calendar. Still, this app here is one of the smartest tools available in the market.

The Calendar is an AI-powered productivity app that is continuously learning from your data, helping you save time and energy as you plan out your day, week, month, and even year. And being AI-powered, the more you use it, the more value it brings to you as it will learn from you every time.

It is also collaborative since you can allow anyone to choose a time and book a meeting with you directly inside the app, releasing you to worry about overbookings, as Calendar will smartly protect you from any panels overlapping.

Habit—Daily Tracker

Ok. Now you have a great and smart calendar app, you can start creating and maintaining good productivity habits to help you achieve your long-term goals.

So, the next step is to find a daily habit tracking app. But why is it essential?

Well, perfection comes from good habits, and with this app, you can create a habit for anything you want to track. The concept behind it is ridiculous simple: If you've done it, check it off.

Smartly tracking your habits can make a difference in your life, and you will start to see the improvements very fast.

First up, you need to decide the habits you need to track, like to read more or start playing chess twice a week; perhaps you're trying to learn a new language.

Once you identify the habit you want to track, you don't even need to open the app to use the widget to manage your habits.

All you have to do is hit the plus sign to add a habit: Give it a name, assign a motivational mantra to keep you on track, and then set the regularity. If you accomplish, you need to swipe to the left, see what you have and haven't done, and check off what you have like that.

The interface is clean and graphic, and there are no distracting bells and whistles.

Things

I know that there are thousands of to-do apps on the market, but Things is beautiful, fast, and easy to use.

The app has been completely rebuilt from the ground up—with a timeless new design, delightful interactions, and powerful new features.

Using Things goes like this: First, you capture tasks in the inbox. Then, you sort your assignments into the appropriate Area of Responsibility (home, work, school) or Project (rebuild the deck, thesis paper).

When it's due, you mark if there is a due date and when you intend to get it done (A specific date/Anytime/Someday).

This makes use of the very powerful Today view. Today, you can see all the tasks you have marked as necessary for today. There should only be three or four, and ideally, by the end of every day, all the functions in your Today view are marked complete.

Some things take several steps to complete but don't require a full-blown project. For those cases, we now have Checklists, which help you break down the finer details of a to-do and outline precisely what's required to get it done.

If you want more to do, you can check the Anytime view. If you want to see what's coming up next, you can check Upcoming. If you're going to see some long-term aspirational tasks, you can check Someday.

If you want to know what you've already gotten done, you can check Logbook. And, of course, you can always check each Area of Responsibility or Project.

Once you've made your plan in the morning, the Today list is your go-to place for all daily activities. Calendar events now display together with your to-dos, giving an outline of your schedule.

Of course, Search, and navigation in Things is now extremely fast, with Quick Find. All you need to do is swipe down in any list and start typing—the name of a project, to-do, or tag—and instantly, you're taken there.

Bonus app: Grammarly

If you like me, write a lot, it can be emails, blog posts, social media posts, or any writing, for sure you can benefit from one of the most exciting productivity apps that I've been using, called Grammarly.

I make massive use of Grammarly every day. It automatically detects grammar, spelling, punctuation, word choice, and style mistakes in my writing. It offers brilliant suggestions to help correct my errors.

Grammarly's browser extension makes it an ever-present entity in every text element I may use, helping me write emails, blog posts, comments, and more.

The app has also added integration with M.S. Word and Google Docs.

Conclusion

So, that's it! These are three(plus bonus) top productivity apps for you to start your journey.

If you want to become more productive in 2020 (who doesn't?), these apps, for sure, will help you, just like they helped me get more done in less time.

By Jair Ribeiro on February 2, 2020.

9 - Illustration vector created by pikisuperstar - www.freepik.com

The European Commission has its Ethical Guide on Artificial Intelligence. Why does it matter?

The European Union hopes that the creation of robust ethical guides will give European technology companies an advantage.

Let's face it... the European Union is not at the forefront of exporting Artificial Intelligence systems, as is the United States and China,

But the continent has been at the forefront of several debates and challenges about ethics with technology and, mainly, with the development of artificial intelligence.

At the beginning of April, precisely on April 8, 2019, the European Commission started the pilot phase of implementing the Ethical Guide for Artificial Intelligence.

The Guide's testing procedures are open to industries, research institutes, and public authorities. It will last until this year, 2020 when a conclusion about the tests will be published.

Building a trustworthy A.I.

The Ethical Guide's development aims to help the Member States create autonomous "trustworthy" systems not to reproduce discrimination and guarantee human autonomy.

When developing systems, the Ethical Guide application will be a differential for the European Union, creating advantages when exporting these products.

Also, the Guide's creators believe that making the Ethical Guide a differential in technological development may encourage other countries and powers of technology and join and develop their Ethical Guides, which would be positive for the contemporary world.

The seven principles

The Guide provides seven principles for the development of an Ethical Artificial Intelligence:

1. Human factor and human control: it means that artificial intelligence systems must be vectors for an egalitarian society, existing at the service of social and fundamental rights, without, however, restricting human autonomy.
2. Robustness and security: an artificial intelligence system considered trustworthy requires that its algorithms are sufficiently safe, reliable, and robust to manage errors and inconsistencies in all phases of a system's life cycle.
3. Respect for privacy and data governance: citizens must know and be fully aware of their data. This data is not used against themselves in a way that generates harm or discrimination.
4. Transparency: the possibility of tracing and retracing the artificial intelligence systems must be ensured.
5. Diversity, non-discrimination, and equity: artificial intelligence systems must consider a whole range of human capacities, skills, and needs; accessibility to this diversity

and plurality must be guaranteed when the system is operable.
6. Social and environmental well-being: A.I. systems should sustain and support positive social developments and reinforce durability and ecological responsibility.
7. Accountability: it is appropriate to apply mechanisms to guarantee human responsibility concerning A.I. systems and their results and subject them to an accountability obligation.

Why do we need an open debate about A.I.?

An open debate about ethical guides has a fundamental value of helping us think of ways to make business, startups, systems, and technologies developed in Europe to follow ethical principles by design.

The European Union hopes that creating robust ethical guides will give European technology companies an advantage in exporting artificial intelligence systems worldwide.

> As Andrus Ansip, Vice-President of the Commission and responsible for the Digital Single Market, said: "The ethical dimension of artificial intelligence is not a luxury tool or an extra. It is only with confidence that our society can fully benefit from technology".

A.I. can bring meaningful benefits to our societies, from helping to diagnose and cure cancer to reduce energy consumption. Still, people must be put in the condition to trust in A.I., know that privacy is respected and that A.I. decisions are not biased.

And what about you? Have you thought about developing an ethical guide for your business?

By Jair Ribeiro on February 3, 2020.

10 - People vector created by pch.vector - www.freepik.com

We should learn how to collaborate with robots before it is too late.

A brief analysis of how Automation impacts our work and how we will thrive in this new era of human+machines.

Much has been said about how automation processes can render a large number of jobs obsolete.

After all, the opportunities for technologies involving robotics and artificial intelligence have grown exponentially. Workers worldwide have been anxious about how this new era of Automation can affect their careers.

The concerns of shrinking jobs during the rise of robotic Automation and A.I. it's a real thing, and it can be contrasted with three main approaches:

The first is related to continuous learning. The second is associated with accessing and analyzing information in the right way. The third is linked to the importance of uniquely human skills.

Let's face it: to be prepared for future jobs, we should be less concerned with choosing a secure job position and devoting ourselves more to the continuous learning of new skills.

Automation and the future of work

Numerous studies try to predict the risk of job losses due to Automation.

For example, Oxford University published a survey in which 47% of jobs in the United States are at risk of being automated. Mckinsey estimates that up to 800 million workers worldwide can be displaced from their jobs because of Automation by 2030.

Some professions, as we know them today, will change dramatically, while others will disappear completely.

As machines take on repetitive tasks and the human work becomes less routine, many jobs will evolve into a new model of work, so-called "superjobs"—jobs that combine parts of different traditional roles into integrated functions, adding significantly human skills to automation technologies such as robotics, cognitive technologies, and A.I.

Entering the era of integration between Humans and Machines

We are coming in the age of a conceptual approach centered on people, which values significantly human skills (such as creativity, empathy, etc.) and emphasizes the need to better understand the highly technological world around us.

To be prepared for future jobs, we must dedicate ourselves to a continuous learning culture based on three main dimensions: Technical, Human, and Data capacity.

Technical capacity: we must understand how machines work and continuous learning about them.

We must ensure that we are trained adequately in the use of new systems and technologies. After all, learning how the technology works and improving our technical capacity is one of the factors that will make us prosper in the future.

To do this, we can use different formats of learning and knowledge sharing systems that massively relies on Group training, Online Education, Webinars, and Lectures by consultants, for example.

In this new era, when we talk about the future of work, we should not feel obliged to adopt any new technology that appears, but be aware of the news in your industry and understand how we can work proactively.

The most important thing is to adopt a culture of innovation and learning present in our daily lives.

Data capacity: we must learn how to analyze and interpret information generated by machines.

Considering this rapid technological advancement scenario and digital transformations, it is essential to avoid becoming hostage to information volume.

Data analysis must be productive, generate useful insights and innovative ideas to improve our company's product and service and internal processes.

We will be called to foster a proactive culture of data management and knowledge sharing in our company. This is nothing more than understanding how the information and statistics developed through new technologies can work for our organization, improve decision-making processes, or provide our team with valuable information that was previously difficult to access.

Human capacity: developing the fundamental human capabilities that machines cannot imitate, or soft skills.

There will be skills that only human beings have in the era of collaboration between humans and machines. Robots will not be capable of imitating the so-called "life-skills."

These skills can be abilities or character traits that are more difficult to learn "in the classroom," but that improves with time and experience.

Among these skills, we have created our capacity to use imagination and original ideas to create something and find new ways to use existing resources.

Also, we have empathy; that is the capacity we demonstrate when we put ourselves in the other's shoes to understand and relate to other people's feelings and emotions.

And, undoubtedly will be very significant to differentiate us from the machines, the ability to find solutions by taking information from a known context and applying it in another context with which it does not necessarily have a connection, something that at least today, machines have a real hard time to do.

These capabilities are very similar to soft skills, that is, the intangible skills of human beings, such as the ability to communicate, empathy or the ability to adapt to changes, for example, as opposed to the hard skills, which designate technical skills, such as knowing how to operate software or a machine.

In the future, employers will consider soft skills as relevant or more relevant than hard skills when hiring since they combine sophisticated knowledge and abilities that are more difficult to be taught.

How to cooperate and collaborate with machines at work

To include technical, data, and human capacity in our work today and in the future, we must undergo a change in culture and practices, rethinking several internal processes in our workplace.

We will need to rethink how we gather data and generate reports of easy access to all, discovering new ways for both humans and robots to thrive collectively.

Ultimately, we aspire that technology can facilitate our day to day activities. Still, neither humans nor machines alone can efficiently support us on the complex task of work in the swiftly evolving world.

We have an excellent opportunity to pursue a healthy balance between humans and machines. We will need adequate frameworks to make it happen in a human-centered workplace and our efforts to build a better society.

By Jair Ribeiro on February 5, 2020.

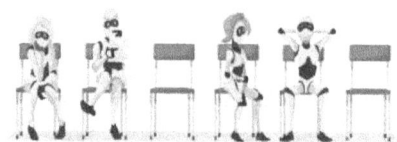

11 - Business vector created by teravector - www.freepik.com

Three facts to consider if you think that A.I. will take your job.

Artificial Intelligence is undeniably changing everything in the job market. but in the end, will it create more opportunities or fewer...

Following examples like HAL, Terminator, Matrix, I Robot, and others, science fiction has risen an endless conflict between humans and machines.

Many questions are being raised with Artificial Intelligence or A.I.'s progress regarding humans and machines' connection.

In particular, one question is recurrent in several conversations I am blessed to attend: which will come in the next several years? After all, will the A.I. eliminate or create tasks?

A.I. and the future of jobs

Concerns regarding jobs are among the most discussed topics when emerging technologies are set on the agenda.

After the media started to constantly display the unrivaled capacities of software and applications powered by Artificial Intelligence, it comes to the heads of tens of thousands of individuals: would robots displace us?

The effects of the improvements that A.I. can deliver are understandably surprising: who'd have believed that machines could one day be in a position to diagnose diseases so accurately, detect suspicious behavior in public surroundings, suggest a solution or content which you're more inclined to eat or even translate legal texts?

It's probably this effect that makes it difficult for humans to see the other side of this coin.

Job Created by Artificial Intelligence

Let's take Google as an example. Twenty and something years ago, it was taking its first steps, and now it's one of the most valuable companies in the world, having a market worth countless billions of dollars.

Along with the consequences of Google's rapid growth, we've seen the creation of countless job openings across the world, jobs that didn't even exist previously.

Marketing and I.T. Professionals started to confront themselves with a new horizon. They had to be trained to adapt to the situation. The result of the google revolution? Even better and more valuable services for millions, maybe billions of people!

Small and medium-sized businesses started to reach more people through internet searches and attract additional clients. Thanks to all those technological innovations, we're ready to do everything on the mobile phone.

In case you did not get it, Google, along with other giants such as Facebook, Netflix, and Amazon, massively utilizes Artificial Intelligence and several other similar technologies to deliver their content to you with only a couple of clicks even before you you know what you want.

The critical point here is that the world changes, and we are invited to learn and adapt in the face of new needs to remain and grow inside the work marketplace.

What's the future between Humanity and the Machines?

> *When people ask me if I believe A.I. will take our jobs, often I used to quote Michaelia Cash, Minister of Labor in Australia, who said: "The future won't be about people competing with machines, it will be about people using machines and doing work that is more interesting and fulfilling."*

I consider this quote a fascinating overview of what's very likely to take place soon.

The workplaces as we know them are changing, and I believe we must acknowledge these changes and be flexible.

The trend is that routine and repetitive jobs will decrease even more. But let's be honest: it has been occurring since the initial industrial Automation. In contrast, new jobs have been created, and new types of work appear.

According to Gartner's estimation, by the end of 2018, roughly two million jobs would be generated in the Artificial Intelligence field by 2023.

Nobody knows for sure what's to come. Still, according to several other historic technological improvements we had before, and what we are living today, A.I. and humanity can work together to offer products, services, and quality of life we have never seen before.

By Jair Ribeiro on February 12, 2020.

12 - Technology vector created by vectorjuice - www.freepik.com

A mini-guide to the E.U.' 's new Artificial Intelligence and Data Regulation plan

European Commission President Ursula von der Leyen wants Europe to have the capability" to create its own choices, based on its values...

The European Union has fresh thoughts about how it could keep up with America and China on A.I.— and it may shape global thinking on how technology is regulated in the procedure.

The European Commission (E.C.)released a bundle of suggestions for Europe's digital future on February, 19th 2020, including a new data plan along with a white paper on A.I.

The proposals would lead to data collected and shared to level the playing field between European companies and competitors from the U.S. and China. The E.C. wants to prevent potential abuses while also building confidence among citizens to reap the technology's benefits.

There will be regulations of cutting-edge Artificial Intelligence applications and the unified European data market, among other topics.

As placed out from an op-ed by E.C. President Ursula von der Leyen, the lait motif is working to provide Europe with the capability to make its own decisions, based on its values, respecting its rules on A.I.

The E.C. did disclose plans to spend almost $21 billion on A.I. and data research programs and the platforms that may eventually allow for the pooling of data envisaged by the Commission. It may sound like an impressive amount of money but isn't going to put the E.U. on a level of investment in the U.S. or China.

> "Given the major impact that A.I. can have on our society and the need to build trust, it is vital that European A.I. is grounded in our values and fundamental rights such as human dignity and privacy protection. Furthermore, the impact of A.I. systems should be considered not only from an individual perspective but also from the perspective of society as a whole."- European Commission President Ursula von der Leyen

But Europe's regulations could, in the process, have a worldwide impact.

Von der Leyen promised to initiate A.I. legislation within 100 days of taking office. The A.I. white paper published today is a set of proposed approaches that could change considerably before becoming law.

It was presented as a roadmap for several rules that need to be adopted in the next years.

However, precisely what do these proposals for what Europe should do mean? Let's have a look.

Artificial Intelligence is the future.

Although A.I. has been labeled critical to economic survival, Europe is perceived as slipping behind the U.S., where development is being led by tech giants, and China, where the central government is leading the push.

With its latest digital strategy, the E.C. wants to encourage more cooperation between the public and private sectors. The plans call for finally creating a single digital market across the continent, a goal the E.C. has been pursuing for years with only limited results.

The E.U. sees enormous potentialities in A.I.: to enhance people's lives through improving effectiveness in areas like healthcare, agriculture, and engineering, and it is a catalyst for economic growth.

However, it knows that American and Chinese tech giants are ahead of the race, and they have to figure out ways to drive investment to grab.

The white paper proposes a way: concentrating on industrial, business, and public sector data that will end up being stored and processed on devices at the edge of the network, rather than in the cloud.

This approach opens up new opportunities for Europe, including a dominant position in the digitized industry and business-to-business software, but a relatively weak spot in consumer platforms that the Artificial Intelligence white newspaper notes.

No facial recognition ban, for now.

A draft of the A.I. paper that leaked in January recommended a provisional ban on facial recognition technologies in public places to provide governments time to define how to use it ethically and safely.

While uses like unlocking a smartphone are seen as relatively safe, the Commission warns that using facial recognition to identify people remotely may pose human rights risks. Such facial recognition use is currently allowed in Europe under very limited exceptions when deemed a serious public interest.

This topic didn't wind up in the published documents this week; instead, the white paper states that the Commission may establish a broad European discussion on the particular conditions, if any, which may require the implementation of facial recognition technologies for identifying people in public areas.

It seems that the E.C. is likely to take regulation within this region seriously, rather than attempting to roll out strategies to satisfy deadlines.

Data is the new oil for Europe.

Finally, Europe started to think about its data strategy seriously, recognizing that only a small number of big tech firms hold a vast amount of the world's data right now, and that is an issue since it can reduce the incentives for European data-driven companies to develop, evolve and innovate at the continent.

So the E.C. lays plans to promote local investment and foster the evolution of local players, including incentivizing data sharing between European companies with an emphasis on the industrial and business data that doesn't run afoul of the E.C.'s strong privacy protections.

Leading these programs creates the European data space, a genuine single market for data that is open to information from anywhere but regulated by European rules and values, including severe personal and consumer data protections.

Promoting the sharing of this data within a structured marketplace may also encourage transparency, which may help Europe more efficiently govern information.

Europe also wants to assess the potential risks of Artificial Intelligence.

The most prominent regulatory principle in the proposal is introducing a compulsory measurement platform for A.I. software E.C. considers high risk, particularly those with notable human rights implications, such as government use of facial recognition or predictive policing algorithms.

The Commission is inspired here by compliance assessment mechanisms currently for many items being placed on the E.C.'s internal market like cars and chemicals.

The proposal lays out two essential standards for what it considers a" high risk" use of A.I. First, is it being set up in an industry where there could be significant risks, like health, energy, or transportation? Secondly, can it be a system that may affect security?

Also, the document suggests a light touch for safe data uses to avoid inhibiting innovation.

However, none of that can be defined. It can have a global impact on how A.I. will be developed and how data will be governed in Europe and large tech companies who have pushed their future vision to get ahead of A. I regulation that may signify a crucial moment.

The A.I. white paper is open for comment until May 19. The Commission can also accept feedback on the data plan.

Conclusion

A.I. is moving at such a pace that we need to regulate it. However, one of the major issues is that when creating A.I., there is considerable input by humans, and humans are naturally prone to having all sorts of bias.

A.I. And machine learning data are only as good as the data you feed it, so the regulation of this needs to start at the earliest stages.

However, regulation will be complicated to follow through with, especially as much of this won't be relevant to the rest of the world."

The E.C. can hardly be counted as one of the data industry's big winners. Its tech firms more often than not playing second fiddle to rivals from the U.S. and China. Still, the E.C., with this initiative, wants to change that.

This is an early stage of a long story, but Europe seems to be prepared to dive into A.I., which could mean the rest of the world soon follows, for the good of humanity, I hope!

By Jair Ribeiro on February 20, 2020.

13 - Technology vector created by vectorjuice - www.freepik.com

Five amazing books about AI that you should be reading

A list with 5 of my favorite AI books will surely inspire you with handy information about several aspects of AI while providing...

I love to read! I'm always reading... and most of all, I love to read about Artificial Intelligence.

I want to share here some of my favorite AI books that surely will inspire you. These are great books for anyone who wants to get serious about AI algorithms and how we humans can relate to this topic today and in the future.

The Book of Why: The New Science of Cause and Effect

The Chancellor's computer science professor and statistics at UCLA and Turing Award winner Judea Pearl and Dana Mackenzie, a Ph.D. mathematician turned science writer, has written for various popular science magazines over the last 20 years, including New Scientist, Scientific American, and Discover.

This book tells the story of science that has changed how we distinguish facts from fiction and has remained under the general public's radar.

"Correlation is not causation." This mantra, chanted by scientists for more than a century, has led to a virtual prohibition on causal talk. What might sound like a reasonable assertion metastasized in the twentieth century into one of science's biggest obstacles, as a legion of researchers became unwilling to claim that one thing could cause another? This all changed with Judea Pearl, whose work on causality was not just a victory for common sense but a revolution in studying the world.

The causal revolution, instigated in this book by Judea Pearl and his colleagues, has cut through a century of confusion and established causality—the study of cause and effect—on a firm scientific basis.

Pearl's work enables us to know not just whether one thing causes another: it lets us explore the world and the worlds that could have been.

It shows us the essence of human thought and the key to artificial intelligence, and if you want to understand it more in-depth, you need to read *The Book of Why*.

How to Create a Mind: The Secret of Human Thought Revealed

Ray Kurzweil, an inventor, and futurist who has published books on health, artificial intelligence, transhumanism, and technological singularity, is arguably today's most influential and often controversial futurist.

In **How to Create a Mind**, Kurzweil presents a provocative exploration of the most critical project in human-machine civilization—reverse-engineering the brain to understand precisely how it works and using that knowledge to create even more intelligent machines.

Kurzweil discusses how the brain functions, how the mind emerges from the brain, and the implications of vastly increasing our intelligence powers in addressing the world's problems. He thoughtfully examines emotional and moral intelligence and the origins of consciousness. The author envisions the radical possibilities of our merging with the intelligent technology we are creating.

Sure to be one of the most widely discussed and debated science books of the year, *How to Create a Mind* is sure to take its place alongside Kurzweil's previous classics, which include *Fantastic Voyage: Live Long Enough to Live Forever* and *The Age of Spiritual Machines*

Life 3.0: Being Human in the Age of Artificial Intelligence

How will Artificial Intelligence affect crime, war, justice, jobs, society, and our very sense of being human? The rise of AI has the potential to transform our future more than any other technology—and there's nobody better qualified or situated to explore that future than Max Tegmark, a Swedish-American physicist, and cosmologist. He is a professor at the Massachusetts Institute of Technology and the Foundational Questions Institute's scientific director.

This book empowers you to join what may be the most important conversation of our time. It doesn't shy away from the full range of viewpoints or the most controversial issues—from superintelligence to meaning, consciousness, and the ultimate physical limits on life in the cosmos.

The Emotion Machine: Commonsense Thinking, Artificial Intelligence, and the Future of the Human Mind

Our minds are working all the time, but we rarely stop to think about how they work.

Marvin Lee Minsky (born August 9, 1927, died January 24, 2016) was an American cognitive scientist in artificial intelligence (AI), co-founder of Massachusetts Institute of Technology's AI laboratory, and author of several texts on AI and philosophy.

The human mind has many different ways to think, said Marvin Minsky, the leading figure in artificial intelligence and computer science. We use these different thinking methods in other circumstances, and some don't even associate with thinking.

"The Emotion Machine" explains how our minds work, how they progress from simple kinds of thought to more complex forms that enable us to reflect on ourselves—what most people refer to as consciousness or self-awareness.

Once we know thinking, we can build machines—artificial intelligence—that can assist with our thinking, machines that can follow the same thinking patterns that we follow, and think as we do. These humanlike thinking machines would also be emotion machines—just as we are.

This is a brilliant book that challenges many ideas about thinking and the mind. It is as insightful and provocative as it is original, the fruit of a lifetime spent thinking about thinking.

Artificial Intelligence: A Modern Approach

The long-anticipated revision of this best-selling book by Stuart Russell, Professor of Computer Science and Smith-Zadeh Professor in Engineering, University of California, Berkeley and Honorary Fellow, Wadham College, Oxford and Peter Norvig, Director of Research at Google, offers the most comprehensive, up-to-date introduction to the theory and practice of artificial intelligence.

The book is as close to exhaustive as is currently available in the field, including in-depth treatments of non-technical learning material while providing an accessible and understandable overview of significant concepts.

Since the 2003 edition, increased coverage has been given to topics such as constraint satisfaction, local search planning methods, multi-agent systems, game theory, statistical natural language processing, and uncertain reasoning over time.

Attention has also been given to providing more detailed descriptions of algorithms for probabilistic inference, fast propositional inference, probabilistic learning approaches including EM, and other topics.

The comprehensive, up-to-date coverage includes a unified view of the field organized around the rational decision-making paradigm.

The authors' approach delivers in-depth coverage of basic and advanced topics. It provides a basic understanding of the frontiers of AI without compromising complexity or depth.

It conveys an in-depth understanding and a clear explanation of such concepts as supervised and unsupervised machine learning. Thus, to the layman, a sense of why there will be no jobs for machine learning supervisors!

It is highly recommended. Intellectually, Artificial Intelligence: A Modern Approach provides both a conceptual artificial intelligence gym and a running track to limber upon. The more you use it, the more you will get from it.

Conclusion

These five masterpieces are not just good AI books, but I would consider them some of the most exciting and thoughtfully structured textbooks I've seen on this subject.

They present useful information about particular aspects of AI while providing fascinating historical background.

I always think that people, especially in the scientific and engineering society, underestimate the importance of simple explanations of difficult concepts, especially concerning new people in the field; these books help make difficult concepts seem difficult!

By Jair Ribeiro on February 23, 2020.

14 - Design vector created by freepik - www.freepik.com

Here is The Vatican's plan for the development of ethical AI.

The Vatican presented a study on bringing more ethics to the development of artificial intelligence for humanity.

Pope Francis launched on last on 28 February 2020 the "Rome Call for AI Ethics" and IBM and Microsoft are among the first signatories.

The document proposed by the Vatican and signed by IBM and Microsoft for the ethical development of artificial intelligence is called "Rome Call for AI Ethics."

The Pontifical Academy for Life, an institution that deals with the ethical and moral implications raised by the latest frontiers of science, such as stem cell research and genetic editing, promoted the document, recognizing that Artificial intelligence (AI) is bringing about considerable changes in the lives of humans and will continue to do so.

AI systems must be conceptualized and realized to assist and preserve human beings and the environment in which they live.

This initial vision must transmute into an engagement to create social and personal conditions that allow both groups and individual members to endeavor to represent themselves where possible thoroughly.

Given the innovative and complex nature of the digital transformation issues, it is essential that all stakeholders work together and that all the needs emerging from AI are represented.

Artificial intelligence is an incredibly promising technology that can help us make the world smarter, healthier, and more prosperous. Provided that, from the outset, it is developed according to human interests and values. The Call for AI Ethics in Rome reminds us that we must think carefully about the needs of those who will benefit from AI and invest significantly in the necessary skills.

Towards an "algorithmic ethics."

The Rome Call for AI Ethics considers three aspects, ethics, education, and rights. The Academy's reflections revolve around the development of AI that is respectful of all human beings. And it ends with the hope of creating an "algorithm-ethics," or an approach of "ethics by design" that should shape each algorithm's development.

The algorithm-ethics is defined through six fundamental principles that should inspire the development of artificial intelligence: transparency, inclusion, responsibility, impartiality, reliability, security, and privacy:

1. **Transparency**: *in principle, AI systems must be explainable;*
2. **Inclusion**: *the needs of all human beings must be taken into consideration so that everyone can benefit and all individuals can be offered the best possible conditions to express themselves and develop;*
3. **Responsibility**: *those who design and deploy the use of AI must proceed with accountability and transparency;*

4. ***Impartiality:*** *do not create or act according to bias, thus safeguarding fairness and human dignity;*
5. ***Reliability****: AI systems must be able to work reliably;*
6. ***Security and privacy:*** *AI systems must work securely and respect the privacy of*

The Call intends to create a movement that expands and involves other subjects: public institutions, NGOs, industries, and groups to produce a direction in the development and use of technologies derived from AI.

The prospect of a good AI

The initiatives promoted by the Vatican and culminated in the Call for AI are part of a more comprehensive scenario of commitment to the ethical development of artificial intelligence.

The European Union recently published a white paper on AI. It launched a public consultation to involve citizens and stakeholders in developing new technologies. In February, the Pentagon also announced new guidelines for the use of artificial intelligence.

Then there are the joint efforts of large industrial groups such as the Partnership on AI, an association to which practically all the big names of IT have joined, from Apple to Amazon, Facebook, and Google, in addition to IBM and Microsoft that conducts research, organizes discussions, shares insights, provides thought leadership, consults with relevant third parties, responds to questions from the public and media, and creates educational material that advances the understanding of AI technologies including machine perception, learning, and automated reasoning.

For an ethical AI development

Now more than ever, we must guarantee an outlook in which AI is developed with a focus not on technology, but rather for the good of humanity and of the environment, of our ordinary and shared home and of its human inhabitants, who are inextricably connected.

A vision in which human beings and nature are at the heart of how digital innovation is developed, supported rather than gradually replaced by technologies that behave like rational actors but are in no way human.

It is time to begin preparing for a more technological future. Machines will have a more critical role in human beings' lives and future. It is clear that technological progress affirms the human race's brilliance and remains dependent on its ethical integrity. [1]

The Vatican focus in the Call for AI is, unsurprisingly, on the human issue: systems, algorithms, etc. they must always be placed in favor of equality between people and promote values aligned with the social doctrine of the Catholic Church: the dignity of the individual, justice, subsidiarity (principle in which the State only acts when the lower spheres are unable to help the citizen) and solidarity. Everything else is secondary.

Conclusion

There is no reason to ban technology, but development and implementation must follow strict rules and regulations to protect people and work for the people, not against them or for the benefit of the few.

Overall, this initiative shows that the Vatican is not alienated from new technologies and seeks to align its values with society's evolution, something it knows how to do very well, with its more than 2,000 years of history.

By Jair Ribeiro on March 3, 2020.

15 - Illustration vector created by stories - www.freepik.com

How I am summarizing the most relevant terms of Artificial Intelligence from Wikipedia, using AI.

This is experimentation on automating Wikipedia articles' summarization with the most relevant definitions, terminologies, and references related to Artificial Intelligence using vectorization (Nltk).

As I started to work on this experimentation, I've managed to select and collect 407 links related to AI, 520 links related to Machine Learning, and 30 links related to Deep Learning (since several terms and definitions are shared between ML and DL).

Suppose you are starting on learning AI or already have some experience in this field. In that case, you can agree that those thousands of articles I've collected represent a considerable amount of data.

It is tough for a human to get reliable and quick insights from such vast volumes of information. Furthermore, a large portion of this data is either redundant or doesn't contain much useful information. The most efficient way to get access to essential parts of the data, without having to sift through redundant and insignificant data, is to summarize the data so that it contains non-redundant and useful information only.

Having this in mind, I decided to build an AI solution that uses automatic text summarization techniques to summarize all the links I've collected on Wikipedia about AI, ML, and DL.

Text summarization is a subdomain of Natural Language Processing (NLP) that deals with extracting summaries from vast chunks of texts. There are two main types of techniques used for text summarization:

NLP-based methods and deep learning-based techniques. In this article, I will use a simple NLP-based technique for text summarization. I will not use any machine learning library in this solution. Instead, I will simply use Python's NLTK library for summarizing the selected Wikipedia articles.

Automatic summarization reduces large text documents to a short set of words or paragraphs that convey the full text's meaning.

An excellent example of the Text Summarization problem is news article summarization, which attempts to produce an abstract from a given article automatically.

Why Text summarization matter?

There are several possible uses of text summarization, like:

- get the full information by spending minimum time from unstructured textual data.
- Enhance the readability of the documents.
- Eliminate redundant, insignificant text, and provide the required information.

- Accelerate the process of researching for information.

Different approaches to Text Summarization

There are two methods used in automatic summarization:

1. The extractive method selects a subset of existing words, phrases, or sentences in the original text to form summaries. In simple terms, we identify the critical sentences or key phrases from the original text and extract only those from the text. Phrases from the original text and extract only those from the text.
2. The abstractive method builds an internal semantic representation. It uses natural language generation techniques to create summaries that resemble humans' ones. This summary may have words that are not present in the original document. Advanced Deep Learning techniques are used to generate a new summary.

There are also two main types of automatic summarization:

- Key-phrase extraction selects individual words or phrases to tag documents.
- Document summarization selects whole sentences to create short paragraph summaries.

My goal here is to make a simple summarizer to flat my AI terminology learning curve; I will be using the Extractive summarization approach.

Steps involved to create the text summary

1) Data collection from the .csv file, then loading URLs using the Urllib library. Data collection from Wikipedia using web scraping(using Urllib library) Fetch the data from .csv file with Title, Url, Website, and Dat, to be used by the Urllib library, which will connect to the page and retrieves the HTML.

I'll be using the urlopen function from the urllib. Request utility to open the web page. Then, I'll use the read function to read the scraped data object.

2) Parsing the URL content of the data(using BeautifulSoup library)

3) Data clean-up like removing special characters, numeric values, stop words, and punctuations.

4) Tokenization—Creation of tokens (Word tokens and Sentence tokens) To split the article_content into a set of sentences, we'll use the built-in method from the nltk library. Import the stop words from the NLTK toolkit and punctuations from the strings library. Stop words are a set of commonly used words in any language. For example, in English, "the," "is," and "and" would easily qualify as stop words. In NLP and text mining applications, stop words are used to eliminate unnecessary words, allowing applications to focus on important words instead.

5) Calculate the word frequency for each word. Word tokenizes the entire text. We have to create the dictionary with key as words and value as the number of times word is repeated, then divide the number of occurrences of all the words by the frequency of the most occurring word

6) Calculate the weighted frequency for each sentence. To evaluate each sentence's score in the text, we'll analyze each term's frequency of occurrence. In this case, we'll be scoring each sentence by its words, that is, adding the frequency of each important word found in the sentence.

7) Create a summary with the top-weighted sentences. Using the nalargest library, get the top-weighted sentences. And later on, join it to get the final summarized text.

Summarizing the articles.

Here I've created a loop that finds the number of articles and runs the function "summarization_links(dfname, index)" for each article present in the dataframe.

How I collected the data?

As mentioned, I decided to summarize articles directly from Wikipedia. When I've started to dig deep into this idea, I realized that Wikipedia is endless; you probably know.

So it would be very hard to work with the classic method: ***research -> copy -> paste*** of the terms and links.. so I used a javascript to collect all links from the Wikipedia web page. here is the code I've used:

```javascript
var x = document.querySelectorAll("a");
var myarray = []
for (var i=0; i<x.length; i++){
var nametext = x[i].textContent;
var cleantext = nametext.replace(/\s+/g, ' ').trim();
var cleanlink = x[i].href;
myarray.push([cleantext,cleanlink]);
};
function make_table() {
 var table =
'<table><thead><th>Name</th><th>Links</th></thead><tbody>';
 for (var i=0; i<myarray.length; i++) {
 table += '<tr><td>'+ myarray[i][0] +
'</td><td>'+myarray[i][1]+'</td></tr>';
 };

var w = window.open("");
w.document.write(table);
}
make_table()
```

When this code runs, it opens a new tab in the browser. It outputs a table containing each hyperlink's text and the link itself, so there is context to what each link is pointing to. After that, I've copied and pasted it into a spreadsheet and performed the necessary data cleaning.

By Jair Ribeiro on June 5, 2020.

16 Infographic vector created by rawpixel.com - www.freepik.com

Summarizing A.I. Articles using A.I.

This is how I'm using Artificial Intelligence to summarize my favorite articles about... Artificial Intelligence and build a weekly newsletter.

I read an endless number of articles every week about A.I., and I believe it is a good thing to collect them and share them with as many people as possible.

But I like the idea of doing this in a "smart" way, so I'm using Artificial Intelligence (what else?) to automate the whole process of data collection, cleaning, and summarization of my favorite articles.

I am sharing a weekly list of the most relevant articles that I read on Medium that I read during my spare time and updated my work.

Article Summarization tool using NLP

I have started an experiment on automating the summarization of the dozen exciting articles about A.I., Machine Learning, and Data Science that I regularly read every week, using vectorization (Nltk) directly from the links to the news.

With the overwhelming amount of new text documents generated daily in different channels, such as news, social media, and tracking systems, automatic text summarization has become essential for me to keep pace and to digest and to understand so much content.

Text summarization aims to extract or generate concise and accurate summaries of a given text document while maintaining key information found within the original text document.

Automatic summarization reduces large text documents to a short set of words or paragraphs that convey the full text's meaning.

An excellent example of the Text Summarization problem is news article summarization, which attempts to produce an abstract from a given article automatically. It concisely represents the latest news as a summary.

Why Text summarization matter?

Overall, automated text summarization technology is also powering business scenarios in a wide range of industry verticals such as media & entertainment, retail, technology, and financial services such as Robo-advisors.

There are several possible uses of text summarization, like:

- get the full information by spending minimum time from unstructured textual data.
- Enhance the readability of the documents.
- Eliminate redundant, insignificant text, and provide the required information.
- Accelerate the process of researching for information.

Different approaches to Text Summarization

Text summarization methods can be either extractive or abstractive:

1. The extractive method selects a subset of existing words, phrases, or sentences in the original text to form summaries. In Simple words, we identify the critical sentences or key phrases from the original text and extract only those from the text. Phrases from the original version, and extract only those from the text.
2. The abstractive method builds an internal semantic representation. It uses natural language generation techniques to create summaries that resemble humans' ones. This summary may have words that are not present in the original document. Advanced Deep Learning techniques are used to generate a new summary.

There are also two main types of automatic summarization:

- Key-phrase extraction selects individual words or phrases to tag documents.
- Document summarization selects whole sentences to create short paragraph summaries.

My initial is to learn how to build a simple summarizer to be used in my weekly A.I. & ML newsletter, so I've decided to start with the Extractive summarization approach.

Steps involved to create the text summary

1. Data collection from the .csv file, then loading URLs using the Urllib library. Data collection from Wikipedia using web scraping(using Urllib library) Fetch the data from .csv file with Title, Url, Website, and Dat, to be used by the Urllib library, which will connect to the page and retrieves the HTML.

I'll be using the urlopen function from the urllib. Request utility to open the web page. Then, I'll use the read function to read the scraped data object.

2) Parsing the URL content of the data(using BeautifulSoup library)

3) Data clean-up like removing special characters, numeric values, stop words, and punctuations.

4) Tokenization—Creation of tokens (Word tokens and Sentence tokens) To split the article_content into a set of sentences, we'll use the built-in method from the nltk library. Import the stop words from the NLTK toolkit and punctuations from the strings library. Stop words are a set of commonly used words in any language. For example, in English, "the," "is," and "and" would easily qualify as stop words. In NLP and text mining applications, stop words are used to eliminate unnecessary words, allowing applications to focus on essential words instead.

5) Calculate the word frequency for each word. Word tokenizes the entire text. We have to create the dictionary with keywords and value as the number of times a word is repeated, then divide the number of occurrences of all the words by the frequency of the most occurring word.

6) Calculate the weighted frequency for each sentence. To evaluate each sentence's score in the text, we'll analyze each term's frequency of occurrence. In this case, we'll be scoring each sentence by its words, that is, adding the frequency of each significant word found in the sentence.

7) Create a summary with the top-weighted sentences Using the nalargest library to get the top-weighted sentences. And later on, join it to get the final summarized text.

Using Transfer Learning for summarizing the articles

Also, I am experimenting with a more advanced summarization technique using a Transfer Learning model called T5.

Transfer learning, where a model is first pre-trained on a data-rich task before being fine-tuned on a downstream task, has emerged as a powerful natural language processing (NLP) on several language understanding tasks.

T5 is a new transformer model from Google trained in an end-to-end manner with text as input and modified text as output.

Combining the insights from using a text-to-text transformer trained on a large text corpus like the "Colossal Clean Crawled Corpus" or C4, the T5 model achieved state-of-the-art results on multiple NLP tasks like summarization, question answering, machine translation, etc.

The T5 model is inspired by the paper Exploring the Limits of Transfer Learning with a Unified Text-to-Text Transformer by Colin Raffel, Noam Shazeer, Adam Roberts, Katherine Lee, Sharan Narang, Michael Matena, Yanqi Zhou, Wei Li, Peter J. Liu.

T5 is an abstractive summarization algorithm. It means rewriting sentences when necessary than just picking up sentences directly from the original text.

By Jair Ribeiro on June 10, 2020.

17 - Technology vector created by vectorjuice - www.freepik.com

Colorizing Black & White images with Deep Learning

Using a CNN Convolutional Neural Network trained on over a million color images to colorize vintage B&W photos.

Since the beginning of photography, Image colorization may have been reserved for those with artistic talent in the past. Still, thanks to Artificial Intelligence, is it possible to colorize black and white images and video with outstanding quality.

One interesting example is the paper *Fully Automatic Video Colorization with Self-Regularization and Diversity,* which refers to one experiment by the Hong Kong University of Science and Technology, which presents a fully automatic method for colorizing black and white films without any human guidance or references.

Typical image colorization methods require some labeled reference. A key innovation of this paper is a novel framework consisting of a colorization network with self-learning techniques.

The researchers used the ranked diversity loss function proposed in a CVPR paper to differentiate different solution modes. They compared their model with two other state-of-the-art fully automatic image colorization.

The new approach was preferred in the percent of comparisons on the DAVIS dataset. The researchers believe their work on self regularization and diversity can inspire future research.

Fully Automatic Video Colorization with Self Regularization and Diversity Could be used in computer vision applications such as visual understanding and object tracking.

The goal with this R&D is to have my personal use. This fully automatic approach can help me generate realistic colorizations of Black & White (B&W) photos and videos.

From the AI point of view, I've decided to follow the approach of an implementation of a CNN ("Convolutional Neural Network") trained on over a million color images, based on a research work developed at the University of California, Berkeley by Richard Zhang, Phillip Isola, and Alexei A. Efros. Colorful Image Colorization.

As explained in the original paper, the authors embraced the problem's underlying uncertainty by posing it as a classification task using class-rebalancing at training time.

I wanted to achieve a plausible color version of the photographs I've selected to use a feed-forward pass on a CNN.

Film and video colorization is not a new technology. Since the beginning of the last century, some fantastic masterpieces of the cinema were painstakingly hand-colored, frame-by-frame, by humans.

Computer-powered colorization started to be used in the 1970s and has been widely used ever since today. Now Deep learning is enabling a fully automatic image colorization. But there has been no corresponding breakthrough in fully automatic video colorization. A key innovation of this paper is a novel framework with self-regularization techniques.

The Deep Learning approach.

The approach for this solution is to implement a feed-forward pass in a CNN ("Convolutional Neural Network") where 1.3 million photos of objects and scenes from ImageNet were decomposed using Lab model and used as an input feature ("L") and classification labels ("a" and "b"). Applying a Deep Learning algorithm (Feed-Forward CNN), final models were generated and are available here: Zhang et al.—Colorful Image Colorization—models.

Source: https://becominghuman.ai/auto-colorization-of-black-and-white-images-using-machine-learning-auto-encoders-technique-a213b47f7339

Working with the LAB

We usually use for digital images the well-known schema RGB. We often code a color photo using the RGB model. Still, unfortunately, the Deep Learning model that I used on this project is the CIE "Lab."

The CIELAB color space or sometimes abbreviated as "Lab's color space) is a color space that got its name from the three additive primary colors, red, green, and blue.

The Lab [aka CIELAB / L*a*b*] completely separates the lightness from color. Think of lightness as some grayscale image. It only has luminosity but no colors. Channel ***L*** is responsible for that lightness (grayscale), and the other two channels ***ab*** are responsible for the colors. As you can see in the images above, the color information is embedded in the ***ab*** channel.

Without looking at ***L,*** you may notice that it is too hard to know what is in the picture from looking at ***ab.*** *T*hat's because of a scientific fact that says 94% of the cells in our eyes determine Lightness (***L***). That leaves only 6% of our receptors to act as sensors for colors (***ab***)

The good news is that, unlike the RGB color model, Lab color space is designed to approximate human vision. It aspires to perceptual uniformity, and its L component closely matches the human perception of lightness. The L component is precisely what is used as input of the AI model, which was trained to estimate the room's remained lightness.

It was also used to train the AI to estimate the room's remaining lightness, which was then used to input the AI.

Remember that you must have Python (version 3.6) and OpenCV (4.0) installed on your machine. I will describe all the process of colorization with Jupiter Notebook in the next part of the article.

This directory contains script to color images.

You can color the images with running the `main.py` script from the root of the project.

```
python -m src.image_colorization.main --method <name of method>
```

Parameter `--model` is optional, if not present `reg_full_model` is default. The models can be chosen from the following list:

- reg_full_model
- reg_full_vgg_model
- reg_part_model
- class_weights_model
- class_wo_weights_model

Conclusion

The original model by Richard Zhang, Phillip Isola, and Alexei A. Efros was trained using "fake" grayscale images (Modern pictures converted to greyscale).

Running this method on real legacy black and white photographs as I did, you can encounter several challenges due to training nature. However, the model can still produce good colorizations, even though the legacy photographs' low-level image quality is quite different from those of the modern-day photos.

With this experimentation, I have shown that colorization with a deep CNN and a well-chosen objective function can come closer to natural colored photos.

AI not only provides a useful graphics output but can also be viewed as a valid tool for image manipulation, as it performs strongly compared to other self-supervised pre-training methods.

Links and Sources:

- richzhang/colorization
  *Richard Zhang, Phillip Isola, Alexei A. Efros. In ECCV, 2016.*github.com
- Mjrovai/Python4DS
  *You can't perform that action at this time. You signed in with another tab or window. You signed out in another tab or...*github.com
- Auto Colorization of Black and White Images using Machine Learning "Auto-encoders" technique
  *In this article, I'm going to show you the main steps to colorize a black and white images using machine learning.*becominghuman.ai
- gitliber/image-colorization
  *This is about an experimental Artificial Intelligent approach for a solution to implement a feed-forward pass in a CNN...*github.com
- Black and white image colorization with OpenCV and Deep Learning—PyImageSearch
  *In this tutorial, you will learn how to colorize black and white images using OpenCV, Deep Learning, and Python. Image...*www.pyimagesearch.com

By Jair Ribeiro on June 25, 2020.

18- Car vector created by vectorjuice - www.freepik.com

The Ethics of AI and Autonomous Vehicles

In a perfect world, AI should be developed to avoid unethical issues, but that may be unlikely since those issues cannot always be predicted. In an automated society, human beings will have the responsibility to support and protect each other more than today.

In the most diverse society sectors, artificial intelligence (AI) is assuming a significant role.

We have no return point, and artificial intelligence will be incorporated into our daily life, professionally or socially, into our future.

With the crescent adoption of technology, some ethical concerns are posed by the notion of "thinking computers" to make decisions like humans.

A practical approach to AI adoption must be researched and examined. This article starts to explore ethical guidelines for the use of intelligent and autonomous systems.

Artificial Intelligence (AI) has been applied widely among us, with potentially great benefits to humanity. Still, at the same time, several concerns regarding AI's unethical use are growing.

In an ideal world, one should configure the AI to avoid unethical tactics, but this could be impractical because it can not be defined beforehand. Research can be used to help regulators, enforcement workers, and others identify problem-sensitive solutions that may be lost in a massive strategy room.

It also indicates that rethinking how AI works in vast strategic spaces could be appropriate to reject unethical outcomes during the learning process explicitly.

From Asimov and beyond

With the development of artificial intelligence, it is increasingly being applied and used in various fields, giving us enormous potential to enhance our environments, change our lives every day, and make the earth more successful.

Simultaneously, as Artificial Intelligence is becoming more mainstream, it is difficult to ignore the ethical and moral experts' questions in robotics. When AI was only an idea present in science fiction works, many questioned its application's limits. Throughout his thesis, the famous writer and philosopher Isaac Asimov created the "Three Laws of Robotics" intending to make the coexistence of human beings and intelligent robots possible:

> First Law: A robot may not injure a human being or, through inaction, allow a human being to come to harm.
>
> Second Law: A robot must obey the orders given by human beings except where such orders would conflict with the First Law.
>
> Third Law: A robot must protect its existence as long as such protection does not conflict with the First or Second Law.

Asimov later added "Law Zero," which is above the others and specifies that:

> A robot may not harm a human being unless he finds a way to prove that ultimately the harm done would benefit humanity in general!

The discussion about how to "talk" Artificial Intelligence in the face of ethical and moral problems is very comprehensive. In a perfect world, AI should be developed to avoid unethical issues, but that may be unlikely since those issues can not be predefined.

Some examples of ethical issues and self-driving vehicle dilemmas

What should a self-driving vehicle do if there is any possibility that it leads to some people's death? The only other option is a cliff, so what is this car doing? For years, philosophers have been discussing a similar moral dilemma. Still, the discussion has a new practical application with the advent of self-driving vehicles expected to become standard on the road in the coming years.

Another much-discussed issue is The Trolley Problem: Imagine that a given autonomous vehicle has six or five passengers, and you must pull a lever to switch to another lane, where only one person will be in the car's path. Would you kill to save five of them?

Other situations can compound this moral imperative. For example, you are on a pedestrian bridge across the road, and you can see a five-person vehicle. There's a huge man behind you, and you know that his weight is enough to stop the car. Is it a moral thing to drive him off the bridge to save the five?

These are not easy-answered questions, I know. When non-philosophical individuals are asked how driverless cars should deal with a situation where either passenger or driver death is imminent, most of them said cars should be designed to prevent passengers' injury. You can read more about it here…

Researchers, led by the Toulouse School of Economics psychologist Jean-François Bonnefon, in their The Moral Machine experiment, an on-line experimental platform designed to explore the moral dilemmas faced by autonomous vehicles.

The platform gathered 40 million decisions in ten languages from millions of people in 233 countries, showing that these differences correlate with modern institutions and deep cultural traits.

When it comes to autonomous vehicles, the experiment presented a series of crash scenarios to more than 900 participants concluding that 75% of the people thought the car would always deviate and destroy the passenger, even if only to save one passerby.

I believe that good activity generates the highest amount of people's happiness. Based on this logic, any action possible to save the maximum number of people should be taken.

When it comes to autonomous vehicles, if a significant number of people are at risk due to an imminent crash, it may be an ethical and rational decision to proceed on its route, even though it means injuring an innocent pedestrian.

Some philosophers who opposed the Trolley issue argue that this approach is too simplistic in front of such a complex problem that required an extensive assessment of the consequences of the action and its moral ownership.

As Helen Frowe, Professor of Practical Philosophy at the University of Stockholm, once said, the autonomous vehicles manufacturers must build vehicles to protect innocent passengers, considering that those in the car should be anyway considered "more" responsible in case of incidents since they were in charge to decide to where the car should go and when, for instance.

What about children and autonomous vehicles?

This is a hot topic, I know. Let me put one more ethical issue here: Let's consider that a self-driving car can hold four passengers or two children in the rear seat. According to my last thought, if all the vehicle passengers are adults, they should avoid a pedestrian.

But what happens if the only passengers inside the car are children going to school in the morning? Can we consider those children as morally responsible in case of a fatal incident using this same logic? Or inverting the parts, should we find ethically acceptable killing one older person in the street to save two children's lives, for instance?

Some people should argue that, in cases where only adults are in the vehicle, we should have to save a vast number of people to make the death of one Passenger ethical, moral, and acceptable.

Are humans predictable?

What if a pedestrian behaves irresponsibly, placing himself in front of an autonomous vehicle with the intent to cause him to deviate, resulting in the death of a passenger? Considering that self-driving vehicles do not judge cyclists and bikers' actions, this legal aspect would be difficult to consider.

Despite hundreds of articles dealing with every little ethical detail, philosophers are far from finding a solution. Is it more unethical, for example, to deliberately divert the car from an alone passerby than actually to do nothing and cause the vehicle to reach a group of people, following a normative notion that morality should optimize satisfaction?

Honestly, I believe that human beings have a responsibility to support and to protect others. Behaviors that intentionally cause injuries or deaths are ethically worse than any action that may prevent them.

Self-driving cars can only meet very rarely in conditions where there are only two courses of action. It is far from unlikely that one day the system will have to choose whether to harm a passerby or a driver.

Of course, cars can only meet very rarely in conditions with only two courses of action. The vehicle can conclude, with 100 % confidence, that any decision will lead to death. But those vehicles may one day have to choose whether to harm a passerby or the driver. We must design new software, systems, and algorithms that consider these questions and make ethical decisions.

Carmakers probably have the most significant responsibility on this topic. Most of the top players in the global market are still "waiting" to express their views on these ethical issues, probably because it seems complicated to find a satisfactory answer given the lack of unanimity on these issues.

The AI ethics issues and our responsible future.

The decision-making of computers leaves room for legal contradictions and errors.

There are several issues in the AI (artificial intelligence) legal controversy. Many countries hurry to develop industrial and military technologies in several ways, in contrast with ethical committees and institutions deputed to put restrictions and regulations in AI to keep it under control.

Specialized panels can review and, in some instances, block proposals by academics and private companies, but a reasonable way to tackle new technologies' threats without hindering technological advancement is Congressional expert consensus.

Conclusion

Humanity is starting to use Artificial Intelligence (AI) extensively, potentially with significant benefits. Yet, questions about AI's unethical use are on the rise.

In a perfect world, AI should be developed to avoid unethical issues, but that may be unlikely since those issues cannot be predefined.

Research on Ethics can help policymakers, compliance officials, and other institutions find problem-sensitive approaches that lack in large areas of our society.

Self-driving cars are expected to become standard on the roads in the coming years. This technology is not immune to regulatory contradictions and ethics issues.

In an automated society, more than today, human beings will have the responsibility to support and protect each other, and AI-powered behaviors that intentionally causes injury or death will be considered ethically worse than any action that might prevent them. We need to start thinking about it now!

By Jair Ribeiro on July 14, 2020.

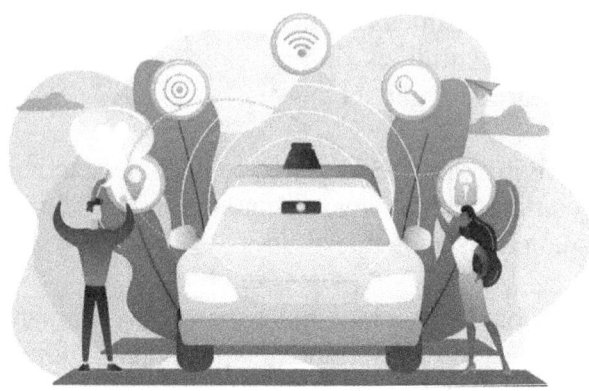

19 - Car vector created by vectorjuice - www.freepik.com

An Introduction to Autonomous Vehicles

Autonomous vehicles have long lived in our imagination since the Jetsons, and if we can imagine, we can do it. The great challenge of mastering gravity has not yet been achieved, but we continue to try on the roads.

Have you ever thought about being in traffic back home and rest in your car by taking your hands off the steering wheel? Or read a book, watch videos while the car is in the chaos of traffic jams?

Well, know that this possibility is already being developed and may soon be available.

I have many tech passions, but autonomous vehicles' development is undoubtedly one of the most intriguing for me among the latest transportation industry trends.

While this innovation seems to usher in a new era in the transportation market, I've been dedicating much time studying and developing a series of "home-made" solutions for autonomous driving that better understand this fantastic technology.

Similarly, it is interesting to know in-depth about how the current self-driving technology works.

Since electronic injection, many automobiles have changed, and other systems have been incorporated into the ECU (central module).

Nowadays, there are airbag control modules, ABS brakes, stability control, autopilot, and immobilizers.

These electronic controllers will need to make a giant leap in quality, efficiency, and safety in autonomous cars.

We will be driving in a better world.

I firmly believe that the impact on traffic safety will be the great legacy of autonomous cars.

Accidents involving cars are the 8th leading cause of death globally, and 95% of accidents are caused by human error.

Bearing in mind that car accidents are the 8th leading cause of death globally, and 95% of the accidents are caused by human error, the expectation is that transport automation will represent a significant reduction in the number of occurrences and, mainly, of victims.

Learning by doing: Home-made autonomous driving systems

The higher the level of automation of a vehicle, the greater the influence of AI and computing on its components. Therefore, I am fascinated to understand the logic that drives these machines attractive, both concerning hardware and software.

Understanding the concepts of electronics and Artificial Intelligence applied to autonomous vehicles has become one of my goals during this period.

I've started with the classic "smart RC cars," coding in python, and adding some necessary sensors to understand how to control small electric motors and servos. Now I'm doing something a little bigger.

I'm working on building an autonomous car for my daughter... well... she is three years old..so it will not be a Tesla Model Y with autopilot. It will still be a "power wheels" ride-on car with level 2 of automation—(see below for Automation Levels). It will use AI and a series of sensors that currently I'm applying to it...

I will write a dedicated article explaining this new project with more details very soon, but before that, let's review together with the basics of this unique technology:

What are autonomous vehicles?

Autonomous vehicles are cars (but also trucks, buses, and others...) where human drivers are not required to take control to safely operate the vehicle, combining sensors and software to control, navigate, and drive the vehicle.

To allow self-driving vehicles to take over the streets, there is a trend by automakers to gradually add technologies that collaborate with drivers both in driving and maintaining the car itself, such as:

- cruise speed control;
- driving and parking assistance;
- braking management;
- obstacle detection system and road users;
- proximity alerts with other vehicles and driving adaptations;
- monitoring of operating conditions;
- speed adjustment according to path conditions.

These are some of the systems already present in specific models. To better understand how integration occurs, let's get to know the technology behind vehicle automation in the following topic.

Currently, there are no legally operational and fully autonomous vehicles in the world. However, partially autonomous vehicles, such as cars and trucks with varying amounts of automation, from assistance for braking to aid changing lanes and parking, with some models even having a certain degree of automatic steering.

Although it is still in its infancy, autonomous driving technology is becoming increasingly common and could radically transform our transport system.

Not every Autonomous vehicle is made equal: the six levels of autonomy.

Different cars are capable of varying levels of autonomy, described on a scale of 0 to 5, and essential to an understanding before we talk about an autonomous vehicle's operation.

The more technological solutions in actuators and sensors the automobile incorporates, the greater its degree of automation. As there are several stages in development, regulations and technical definitions also need to adapt.

For this reason, the Society of Automobile Engineers (SAE) created a classification to differentiate vehicles according to their degree of automation, making it easier for consumers and maintenance professionals to identify the models. The following five levels have been determined:

Level 0: humans control all significant systems

Level 1: specific systems, such as cruise control or automatic braking, can be controlled by the vehicle, one at a time.

Level 2: the vehicle offers at least two simultaneous automatic functions, such as acceleration and steering, but requires human beings for safe operation

Level 3: the vehicle can manage all critical safety functions under certain conditions, but the driver must take over when alerted

Level 4: the vehicle is fully autonomous in some driving scenarios, although not all.

Level 5: the vehicle is fully capable of autonomy in all situations

As we can see, while some models are already produced in series with some level of automation, other prototypes and projects are being touched by automakers and technology companies.

So it is interesting that mechanics and other maintenance professionals start to prepare for the near future. In the next topic, we'll talk about it.

With different degrees of autonomy, autonomous vehicles become popular in the coming years and help make our daily lives much more comfortable.

Based on automakers and technology estimates, level 4 autonomous cars may already be sold in the 2–3 coming years.

How do autonomous vehicles work in practice?

The development of autonomous vehicles is at an advanced stage. Artificial Intelligence today, using computer vision and other methods, allows the vehicles to differentiate the types of obstacles and situations on the roads so as not only to react according to pre-established parameters but also to learn eventualities.

With connectivity, on-board computers exchange information with each other. Thus, an unforeseen occurrence with one of the automobiles will help everyone learn to deal with identical circumstances. For this, automobiles will need more powerful computerized central than current electronic modules (ECUs) and OBD (On-Board Diagnostic) diagnostic systems.

Most self-driving systems create and maintain an internal map of their surroundings based on information obtained from a wide range of sensors, such as radar. Some autonomous vehicles use laser beams, along with other sensors, to build the internal map. Others use radar, high-powered cameras, and sonar, and maps loaded on their systems for operation.

The software then processes the information obtained in real-time, traces a path, and issues instructions to the vehicle's actuators that control acceleration, braking, and steering. Rules, algorithms to avoid obstacles, predictive modeling, and discrimination between objects (that is, knowing the difference between a bicycle and a motorcycle) help the software follow traffic rules and navigate obstacles.

While partly autonomous vehicles (levels 0,1,2, and 4) may require a human driver to intervene if the system encounters conflicts, the future fully autonomous vehicles (Level 5) may not even offer a steering wheel.

Autonomous vehicles can be further distinguished as connected or not, indicating whether they can communicate with other vehicles and with the city's infrastructure, such as the next generation of traffic lights and traffic management in cities.

The technology behind autonomous cars?

The idea for autonomous vehicles is to reach the point where the driver is not required, allowing all occupants to enjoy the trip. For this, technology needs to evolve to a level of reliable security.

Several innovations are being developed, improved, and integrated into autonomous cars to make this scenario possible. Let me share here some of the most important ones:

The Hardware

They are responsible for detecting environmental characteristics and passing this data to the on-board computer. Currently, the most used are cameras, radars, sonars, and LIDARs.

Stereoscopic camera

Also called a stereo camera, it is a device that uses two or more lenses to create frames from different perspectives. In this way, you can get a sense of depth (3D), simulating human vision.

Infrared camera

The infrared camera allows accurate viewing in low or no light environments. This equipment can identify objects by the temperature variation using sensors, capturing their infrared radiation, invisible to the naked eye.

Radar

A radar emits radio waves in a specific direction, which reverberates through obstacles. By measuring the speed and intensity of this return, you can have notions of size and distance.

Sonar

Sonar works similarly to radar. Instead of radio waves, the difference is that it uses sound waves, inaudible to the human ear.

LIDAR

LIDAR also follows the logic of the two previous devices. However, laser pulses, which form thousands of luminous points, are used to scan the environment. In addition to having a faster signal, LIDAR allows you to cover a wider area, 360 °, and greater precision.

ESC (Electronic Stability Control)

Electronic Stability Control is the same technology used in several models, including in Brazil. It is responsible for calculating and making corrections in driving according to each wheel's speed, inclination, and yaw in autonomous cars.

iBooster

The vacuum electromechanical servo brake, called iBooster, can generate controlled pressure on the brakes in less than 120 milliseconds. This is three times faster than conventional brake systems, making the vehicle safer in emergency braking.

GPS, speedometer, and odometer

For the vehicle to guide itself through the cities, it is necessary to equip it with updated maps and control its location. Therefore, it will use GPS equipment integrated with the speedometer and odometer. Thus, the computer can calculate its position even in the absence of the satellite.

The software

While the autonomous car's hardware components allow the vehicle to perform such functions as seeing, talking and moving, the software is like the brain that processes environment information to know what action to take-whether to drive, stop, slow, etc. Three systems can categorize autonomous vehicle software: perception, planning, and control.

Perception

The perception system refers to the autonomous vehicle's ability to recognize what necessary information enters through sensors or V2V components. It helps the car to understand from a given frame whether an entity is another vehicle, a person, or something else entirely.

This mechanism is similar to how our brains translate knowledge into meaning by sight. Our eye photoreceptors (sensors) detect light waves from the atmosphere and transform light waves into electrochemical signals. Neuron networks transfer these electrochemical signals back to the brain's visual cortex, where our brain understands what these electrochemical signals; thus, our.

Thus, our brain understands whether a specific light pattern hitting our retina represents a chair, plant, or another individual.

Planning

The planning system refers to the autonomous vehicle's ability to make certain decisions to achieve higher-order goals. And the autonomous vehicle knows what to do in a situation—whether to pause, move, slow down, etc.

The planning system works by integrating collected information about the environment (i.e., from sensors and V2X components) with defined policies and expertise about how to move in the environment (e.g., do not drive over pedestrians, slow down when approaching a stop sign, etc.) so that the car can decide what action to take (e.g., overtake another vehicle, how to reach the destination, etc.).

As the autonomous car planning system, the human brain's frontal processes lobe processes allow us to think and make decisions like what to wear in the morning or what to do for fun on the weekend.

Control

The control system concerns the process of translating the planning system's objectives and priorities into actions. Here, the control device informs the necessary inputs' he necessary inputs' hardware (actuator leading to the desired motions.

For example, an autonomous vehicle, realizing it should slow when entering a red light, converts this awareness into braking practice.

In humans, the cerebellum processes play an analogous function. The cerebellum is responsible for the essential motor control function.

It encourages us to chew when the desired goal is to eat.

Artificial Intelligence and Connectivity

Artificial Intelligence would be responsible for collecting all sensor signals, internal and external, monitoring driving, notifying the owner of maintenance needs, making minor device changes and improvements, and learning from failure.

Connectivity with other autonomous cars can exchange experiences and solutions.

Such technologies make vehicle automation possible.

Vehicles that can see the world.

It's easy to fall into the illusion that autonomous vehicles perceive their world in the same manner as we do—by "seeing" the environment using stereoscopic vision to assess the vehicle's relative locations and its surrounding elements, but this is not the case.

An Autonomous vehicle's vision system often consists of several high-resolution cameras covering the vehicle's front, sides, and back. Unlike lidar, cameras can detect color, which is useful for spotting traffic lights, construction zone signs, and emergency vehicle lights.

These cameras are designed to work in daylight and low-light situations. Still, like any other camera, the resolution drops as the light decreases.

When the car starts its journey, the software interprets hundreds of objects in different ways, such as cyclists or pedestrians, and reads traffic lights and signs.

Do you foresee the unusual

What if an object falls from a truck in front of you? What if a cyclist hidden between two cars wins the street as if he came out of nowhere? Google says it has been training the autonomous vehicle to handle unexpected situations. In such cases, the car's decision is generally to slow down and wait until more information is captured about what is around.

The Autonomous driving revolution

The expectation is that autonomous Vehicles level of 4 or 5 (full autonomy) become a primary global market in 2030, moving somewhere around $ 60 billion, according to statistics data platform Statista.

By 2035, North America should have 29% of the world's autonomous fleet, with China and Western Europe having 24% and 20% of the autonomous car fleet.

The technology of autonomous vehicles is very complex. In the UK, 73% of all cars are expected to have some level of range (levels 1 to 3) before fully autonomous vehicles start to enter the market, predicted in 2025.

One reason behind this is the lack of consistent high-speed internet connection to allow self-driving vehicles to communicate and gather information about driving conditions and congestion or possible obstacles that block the road.

Another reason is that some vehicles require extremely detailed maps to navigate safely.

Automakers and technology companies are investing heavily in the autonomous vehicle market. No wonder. Research indicates that 85% of adults in the United States would feel safe sitting behind the wheel of a driverless vehicle.

Conclusion

This new wave of vehicles will have a monumental impact on our lives. It is already redesigning the transportation chain. This is because the autonomous (and the "semi") do not only need exhaust, engine, and gear to run. That's the least of it.

They embed "under the hood" a myriad of electronic devices, such as sensors, chips, radars, connectivity points, cameras, and, to sum up, all sorts of algorithms.

In other words, vehicles are becoming computers with tires.

All of this requires that automakers have new suppliers—or that old ones shake off. Hence the origin of all change. There is also much money involved in this transformation. I will write about it in another article.

Since the Jetsons, insinuations in the future vehicles would be smart enough to take us wherever we wanted, even more so by air and not stuck to the roads.

However, the great challenge of mastering gravity has not yet been achieved, but we continue to try on the roads.. and the times seem to be mature now.

By Jair Ribeiro on August 28, 2020.

20 - Car vector created by vectorjuice - www.freepik.comHow

Autonomous Vehicles will redefine the concept of mobility

They are already among us and will transform the entire automotive industry.

The technology behind Autonomous vehicles can surprise you. These vehicles are characterized by not having to deal with human limitations, such as tiredness and inattention.

To the delight of many, these machines can park alone, and they do not drive drunk or speak on the phone while driving, like many humans that we know.

It is known that human failures cause 94% of traffic accidents, and this innovation is mainly developed to save lives, reducing the fatalities consistently.

According to a study from 2015 by the National Highway Traffic Safety Administration (NHTSA), traffic accidents are the most significant cause of death of young people between 15 and 29 years globally, overcoming the victims of AIDS, flu, and dengue together, according to the World Health Organization (WHO).

Also, we always complain that we have little time. The days are busy, and we have several obligations to fulfill. And we spend, on average, 40 days a year stuck in traffic in the cities.

Have you ever thought how much we would earn if we didn't have to worry about driving, with the vehicle working for us during that time?

It is estimated that autonomous cars can create a 7 trillion dollar market! It is called the "passenger economy" since everyone will do several other activities without paying attention.

Beauty salons, dinners, and health clinics are services that can be performed inside the vehicles.

Autonomy in vehicles will transform the entire automotive sector, from the way traffic is organized through vehicle engineering and the parts and components industry to the mechanics.

A disruptive wave: Transportation as a Service.

Some of us still remember when digital photography began to emerge as an option to analog. Kodak carried out studies that proved accurate, indicating that the new technique would dominate the market in 20 years. Despite having time for this, the company wrinkled its nose and did not modernize. Currently, the giant has shrunk and survives only as a patent laboratory.

To avoid the same mistake, almost all automakers are running after losing the future market, planning to deliver vehicles capable of automation. The development of entirely autonomous vehicles by technology companies has already spurred new hires at major brands.

In the article "How Uber's autonomous car will destroy 10 million jobs and redefine the economy in 2025," Zack Kander argues that the entire industry will be reinvented in the next decade. How can we not agree on that?

When the autonomous vehicle technology will be already well developed and a majority in the streets, the tendency is that the market changes will start to consolidate. The entire transportation industry business model is going to change. Vehicles will stop being a consumer product to become a service.

Today, for the majority of the time, a car, for example, is used only for a short time, but the cost to maintain it is not so low. According to a report of GlobalFleet, a car cost varies up to €344 per month across Europe.

Self-driving vehicles will drastically reduce the transportation costs offered by companies when human drivers will not be required. The cheaper service could make the maxim of having the ownership of a vehicle no longer attractive.

We can consider that vehicles in the future will be standardized, electric, and autonomous and will be mainly owned by fleet companies. It means that it will be difficult for the current production logic—with millions of cars manufactured every day—to survive this disruption.

According to Statista, the global auto industry is expected to sell 59.6 million automobiles in 2020. Well, at least it was like that before COVID-19.

Due to the coronavirus pandemic, the sector is projected to experience a downward trend in a slowing global economy. Before the pandemic hit, it has been estimated that international car sales were on track to reach 80 million in 2019; Anyway, the global automotive industry is estimated to reach 1,14,250 thousand units by 2024.

The importance of this revolution for the global economy is evident: Workshops, tire shops, insurance companies, dealerships, professionals (such as drivers, taxi drivers, services such as garages, rent, etc.) will be heavily disrupted. Also, of course, to the assembly lines themselves.

With more significant autonomy comes more significant challenges.

One of the most significant differences between autonomous cars is their level of connectivity. They will be connected to the internet all the time. Because of this, much of what we do today by cell phones or computers will be done during a journey, using the vehicle itself.

On the other hand, Autonomous Vehicles may become a preferred target of virtual attacks, aiming to steal data or compromise systems. Thus, problems with viruses and other malicious programs will be more constant. Are we ready for this?

Autonomous vehicles and the new customer experience.

As the plan's progress in launching commercial autonomous vehicle services by most automakers proceeds, designers and product owners recognize that more than the development of the technology itself, focusing on the customer experience will be essential for future business success.

There are no easy shortcuts when it comes to launching such complex technology as autonomous vehicle services are. The auto industry has faced the most challenging thing since people switched from horses to cars.

The AV adoption heavily depends on people to trust the technology enough to get into the vehicle and then love the user experience to come back.

In the coming years, autonomous vehicles have enormous potential to expand access to transportation, goods, and jobs in several cities.

The best way to do this is to create transportation services that offer customer-centric experiences at every step of the journey.

The challenges of launching a new mobility service with large-scale operations will involve fundamental consumer behavior changes: the exchange of vehicle ownership for vehicle sharing.

This cultural change will require transportation companies to manage their autonomous fleets based on intelligent technology and a high utilization rate. Another critical aspect will be creating reliable and efficient autonomous vehicle services that meet customer expectations regarding vehicle cleaning, maintenance, recovery, and durability to earn their loyalty.

It will also be necessary to know how to program the service's expansion, defining where and how it will be launched to improve the customer experience before gaining a local, national, and global scale.

Our daily relationship with automobiles is about to change. Our user experience will probably be based on more efficiency and safety because these autonomous systems will replace much of the interactions between humans and vehicles. Thus, fewer human errors, and more accumulated knowledge and experience, wear, breaks, and failures tend to be less recurrent, impacting the entire automotive sector.

The impact of Autonomous vehicles on automakers

Since cars were invented more than 100 years ago, the automakers' business model has hardly changed.

In the last years, engines have become more powerful and efficient. The design has become more attractive, new safety and convenience items were added, but, in essence, cars are still quite similar to those cars sold in the early 20th century.

Now, with autonomous vehicles, will this change? It is going to change a lot!

For example, in general, a factory produces a car. As we know today, most of the vehicles pass it on to a final consumer, most of the time, with a dealership's intermediation. Then, the following year, the automaker launches a more modern model and resumes its sales cycle. This process is about to change heavily with autonomous cars.

And the way they generate money for automakers, too.

The automobile industry is also known as the *industry of industries*, "not only for its size and economic importance but also for all the impact and influence on numerous areas of the economy such as commodities, energy, credit, technology, etc. This industry will changes a lot in the next 10 or 20 years with autonomous car technology development.

The idea of having a car that drives itself and that can receive *updates* during its life cycle turns everything we know about the automotive industry upside down.

About maintenance of Autonomous vehicles

With full autonomy, driving will be more efficient, reducing the need for maintenance per kilometer traveled.

There will be less sudden braking, sprinting, unnecessary accelerations, and impacts with holes and obstacles and fewer accidents. After all, sensors and actuators are faster and more accurate than humans.

However, the wear and tear of parts such as shock absorbers, brake pads and discs, tires, cushions and stops, and the end of the useful life of items such as filters and fluids, will practically not change— they will only reach their most efficient stage.

If, or should I say, when autonomous vehicles adopt electric motorization, electrical and electronic systems will replace all parts of fuel injection, combustion, cooling, and exhausting of gases.

Electric vehicles need different maintenance, as they do not deal with explosions and high temperatures.

But does this mean that the car services and part suppliers will end? Probably not.

Someone will still need to manufacture all of these parts and get them to work together. Autonomous Vehicles will always ask for maintenance services, replaced components, and technical and mechanical improvements will still be in development. Activities such as design will also continue to advance. But will all this continue to generate money as before, in a future of autonomous and shared cars?

We can observe this phenomenon observing millennials; many of them no longer buy cars, especially in big cities. If you live in a city like New York, you probably don't want to pay to park your car or deal with other problems when you have a car. I live in Wroclaw, in Poland... and guess... I do not have a car... funny enough for one who works on disruptive technology as a worldwide leader in the transportation industry.

The automakers' business model (making, selling, starting over) may lose some sense when people stop buying their vehicles so often. Still, no one is better positioned than automakers today to lead the (r)evolution of the future vehicles' manufacture.

What changes with the arrival of autonomous vehicles on the transport industry chain?

The popularization of autonomous vehicles will revolutionize the transportation industry in practically all areas.

Here you have some activities that may be affected:

Spare parts market

Even with fewer accidents and more efficient driving, the spare parts market will always be growing. New vehicles will continue to arrive from the factories, adding to an existing fleet.

What will change, without a doubt, is the complexity of the types of parts and their applications, as the diversity of vehicle models tends to increase, requiring the use of tools that are also more complex, in addition to other technology items, so that the mechanic will be able to carry out repairs.

Also, electrical and electronic items like cameras, Lidars, and other sensors will indeed become more relevant in this market.

Service stations

In addition to servicing vehicles with combustion engines (which will continue to run for a long time), the service stations will need to adapt to supply electric recharging services. They will also need to adjust to receiving autonomous vehicles, with new forms of service and payment.

Thus, the diversity of services and products sold will increase, despite the decrease in demand for minor repairs and overhauls, such as changing oil or filters, at gas stations.

Mechanical workshops

Workshops will change to carry out the maintenance required by new technologies, with appropriate tools and trained professionals, if they want to catch the opportunity of expanding their market.

Besides repairs on traditional vehicles, other services will become more requested, such as the review of embedded electronics sensors and electrical components.

Vehicle trades

There is a strong tendency for autonomous vehicles to be shared, accelerating the process of transforming vehicles into services instead of being consumer goods, as they are today. Thus, the number of owners will decrease, especially in large cities, concentrating on fleets.

In this scenario, small and medium auto sales stores will suffer since acquisitions will take place on a larger scale, probably in direct negotiations with dealerships or automakers.

Conclusion

With these simple imagination exercises, we can see that the industry will be touched as a whole. Some actions are already taken regarding auto repair shops that can facilitate this transition to the market dominated by independent automobiles. Some others will take time and depend on how fast the market will react to autonomous vehicles. Autonomous vehicles are already a reality at their most basic levels of automation.

Automakers and technology companies have invested a lot to get their fully autonomous vehicle designs off the ground. Therefore, it is the best moment to follow this technology's evolution and to be prepared for the changes it will bring to our society.

Was this article helpful to you? Want to know more about autonomous driving technology? Let me know.

By Jair Ribeiro on September 4, 2020.

21 - Background vector created by rawpixel.com - www.freepik.com

Fourteen inspiring and influential women who defy the gender gap in Data Science!

Just 15% of today's scientists are women. Like most STEM fields, data science has a daunting gender diversity problem. We need to do something about it.

Last week, I was having a "career conversation" with my 13 years old daughter. Surprisingly, she has shown me a great interest in Artificial Intelligence and Data. I would not define her as a nerd, but she always got excellent STEM grades in general.

As a father, I always try to encourage her, mentor her, and support her in the long way to come when it comes to her career and her decision, so I will do what it takes to give her all the possible opportunities to do her experiments until she finds out what it does for her in the life… but I must confess… there many things we need to do today to support the career choices of female students.

There is a massive lack of diversity: as few as 15% of data scientists today are women. And the lack of diversity is a serious issue. A.I. algorithms are biased, so building them requires a team that includes a wide range of views and experiences.

Achieving diversity of approaches and viewpoints is critical in building efficient data science teams. Machine learning algorithms can occasionally "see" patterns that lead to spurious, biased, or even dangerous conclusions. It takes a diverse group to ensure that bias-prone models produce accurate, balanced results. Building such algorithms can be an art as science.

But as explained by one study of BCG—Boston Consulting Group, data science, like most STEM fields, has a daunting problem of gender diversity.

While women make up about 55% of university graduates across countries on average, they account for just over one-third of STEM degrees. (See Exhibit 1.) Only two-thirds of this valuable talent pool embark on a STEM-related career, such as engineering, analytics, software development, and even less on a data science career. According to different surveys, only about 15% to 22% of all data science professionals are women.

Also, While data scientist is the most promising job position for the next years, according to Harnham's Diversity Report 2020–2021, females occupy only 18% of today's data science roles, and 11% of data teams don't have any women at all.

A career disconnection

According to Girls Who Code, during the last 30 years, female computer science graduates dropped from 37% to 18%. Girls and STEM-related career paths disconnect before college. 74% of middle school girls are interested in STEM topics and careers. Still, only 0.4% of high school girls choose computer science for a major college.

Girls whom Code presents a few reasons why this is the case:

Inherent Bias: The narrative usually defines a "techie" individual as a geeky computer-tinkering male. This idea seems to influence girls to turn away from computer science as an option for future studies.

No Early Exposure: The lack of a proper introduction to computing skills, such as coding or programming, at an early stage can cause a disconnection between the initial interest in STEM-related topics and the choice of other career paths.

Confidence Issues: When girls do not have confidence in skills and potential success, they limit themselves. Internalized stereotypes can force women to feel they don't have the "right" knowledge or skillset for success in the field.

We need more women in Data Science.

Women are more aware of the risks that are a plus in big data. Women tend to excel in communication, team nurturing, and problem-solving. They have the natural talent to ask the right questions and listen to all the data's answers.

Gender diversity can appeal to new customer bases, opening up untapped business opportunities for companies.

The concept of a data science team is relatively new, so companies need to hire leaders who can work horizontally and listen to new ideas.

However, closing the technology gender gap should be about reaching a certain women-to-men ratio within the field. While recruitment remains an issue, the focus should be on empowering women in the area, recognizing their achievements, and reminding girls and women interested in or entering the field that skills are not gender-based.

Unfortunately, the gender gap in technology and data science remains, but bringing more women in isn't necessarily a push. Instead, it's a movement to raise awareness of the current situation, highlight growth opportunities, and highlight the benefits of gender diversity.

Women bring a lot, as they always have, to the technology table. But the support and recognition for what they get will continue to close the gender gap at a slow, steady pace...

An inspiring list of women in Data Science

As we saw, women are underrepresented in STEM science, technology, engineering, and math fields. However, being a father of three amazing daughters and having built my career in the tech industry, I will keep doing my best to support the increase of the number of women in Tech, particularly in Data Science.

During my conversations with my daughter, I've started to mention some inspiring women that I've been following on the internet due to their exciting contribution to the community and their inspiring careers.

Here you have an inspiring list of women's profiles in Tech that for sure can inspire many other women in pursuit of a career in A.I., Analytics, Big Data, Data Science, Machine Learning, and Robotics as well.

Fei-Fei Li

Associate Professor at the C.S. Dept. at Stanford, Director of the Stanford Artificial Intelligence Lab and the Stanford Vision Lab.

Dr. Fei-Fei Li is the inaugural Sequoia Professor in the Computer Science Department at Stanford University and was the Director of Stanford's A.I. Lab from 2013 to 2018.

Dr. Fei-Fei Li is the inventor of ImageNet and the ImageNet Challenge, a critical large-scale dataset and benchmarking effort. Dr. Li is also the author of more than 200 scientific articles in top-tier journals and conferences, with technical contributions as a leading national voice for advocating diversity in STEM and A.I.

Dr. Li is co-founder and chairperson of the national non-profit AI4ALL to increase inclusion and diversity in A.I. education.

Dr. Li was Vice President at Google and Chief Scientist of AI/ML at Google Cloud during her sabbatical.

She works with the world's most brilliant students and colleagues to develop smart algorithms that allow computers and robots to see and think and perform cognitive and neuroimaging experiments to discover how brains see and think.

LinkedIn: https://www.linkedin.com/in/fei-fei-li-4541247/

Twitter: Fei-Fei Li

Follow her on Medium: Fei-Fei Li

Cassie Kozyrkov

Chief Decision Scientist at Google, Inc.

Cassie is Google's data scientist and pioneer in democratizing decision intelligence and secure, useful A.I.

Cassie was born in Saint Petersburg, Russia, and grew up in Port Elizabeth, South Africa. She studied economics and mathematical statistics at Nelson Mandela University. Kozyrkov worked as a PM and researcher at the University of Chicago. In 2016, she was promoted to Chief Data Scientist in the CTO Office at Google and later to Chief Decision Scientist in 2017. She focuses on Google is on applied A.I. and data science process architecture. She has been a keynote speaker at Web Summit, the world's largest technology event. She appeared on the cover of the Forbes AI data science issue. She was named the LinkedIn #1 Top Voice in Data Science and Analytics in 2019.

Twitter: @quaesita

Follow her on Medium: Cassie Kozyrkov

LinkedIn: https://www.linkedin.com/in/cassie-kozyrkov-9531919/

Allie Miller

US Head of A.I. Business Development, Startups, and Venture Capital at AWS, Forbes AI Innovator of the Year.
Allie Miller is Amazon's U.S. Head of AI Market Growth for Startups and Venture Capital (AWS), advancing the world's largest A.I. firms. Previously, Allie was IBM's youngest woman ever to create an artificial intelligence device—spearheading large-scale product creation through computer vision, conversation, data, and regulation. She was IBM's youngest woman to create an artificial intelligence product. Forbes named Allie and A.I. Summit as 2019's "A.I. Innovator of the Year" Allie is also the founder of The A.I. Pipeline to create more significant equity in ML, a national ambassador at the American Association for the Advancement of Science (AAAS), and an ambassador to 10,000-person Advancing Women in Product organization. In three national innovation contests, she won the Grand Prize and spoke about A.I. worldwide. She holds the Wharton School double-major MBA and Dartmouth College B.A. in Cognitive Science ...

Twitter: https://twitter.com/alliekmiller

LinkedIn: https://www.linkedin.com/in/alliekmiller/

Follow her on Medium: Allie Miller

Elizabeth M. Adams

Keynote Speaker | U.N. Key Constituent of Roundtable 3C on Artificial Intelligence| Stanford University Fellow: Race & Technology | Civic Tech

Elizabeth Adams is a technology integrator, working at the intersection of Cyber Security, A.I. Ethics, and A.I. Governance. She also passionately teaches, advice, consults, speaks, and writes on the critical subjects within Diversity & Inclusion in Artificial Intelligence. She is a member of the IEEE Global Initiative on Ethics of Autonomous and Intelligent Systems, helping to build global standards for A.I. Nudging & Emotion A.I. and an appointed member of a Racial Equity Community Advisory Committee for the City of Minneapolis. She's refined her leadership acumen in tech design by leading various technology initiatives in the Washington D.C. metro area.

She has refined her leadership acumen in tech design over the past 20 years by heading numerous technology projects in the Washington D.C. Tube region. Returning to Minnesota's home state, she remains committed to embedding ethics and human-centeredness into artificial intelligence systems and still takes time to fulfill her passion for lifting other women in Tech.

LinkedIn: https://www.linkedin.com/in/lizadams/

Tamara McCleary

CEO at Thulium
Tamara McCleary is the CEO of Thulium.

She harnesses artificial intelligence, machine learning, data, and analytics to drive smart social in the B2B and enterprise space.

She has been featured multiple times in Forbes for her pioneering influencer marketing strategies on social media.

She is a technology futurist, host of podcasts, TechUnknown and SAP Industries Live, keynote speaker, and unique advisor to leading tech companies.

LinkedIn: https://www.linkedin.com/in/tamaramccleary/

Twitter: @TamaraMcCleary

Carla Gentry

Data Scientist
Carla was daunting to be a single mother with two sons, but she never backed a challenge.

Eager to learn and develop, she joined Chattanooga's Tennessee University in spring 1993. She worked for UTC in the

Developmental Math Lab, helping students in all stages of mathematics.

After graduating from UTC with a dual major, Applied Mathematics and Economics, in 1998, she moved to Chicago to begin her analytics career.

Carla has over 19 years of experience working with several Fortune 100 and 500 businesses.

She serves as a liaison between the I.T. department and executive staff, taking complex databases, deciphering market needs, and returning with intelligence that quantifies investment, benefit, and patterns.

LinkedIn: https://www.linkedin.com/in/datanerd13

Twitter: @data_nerd

Follow her on Medium: Carla Gentry

Danielle Belgrave

Principal Researcher at Microsoft Research.

Danielle Belgrave is Principal Research Manager at Microsoft. She works on integrating expert scientific knowledge to develop probabilistic machine learning models. Most of her work has focused on developing models to understand disease heterogeneity in the context of asthma.

LinkedIn: https://www.linkedin.com/in/danielle-belgrave-704157107/

Twitter: https://twitter.com/DaniCMBelg

Kristen Kehrer

Machine Learning Storyteller, Founder of Data Moves Me, and Data Science Instructor at UC Berkeley Extension

8 Global 2018 LinkedIn Top Voice—Data Science & Analytics. Kristen is currently a Data Science teacher at UC Berkeley Extension, Faculty / SME at Emeritus Management Institute, and Data Moves Me Founder, LLC. Kristen has provided creative and actionable machine learning solutions across various sectors, including infrastructure, healthcare, and eCommerce, since 2010. Kristen holds an M.S. at Worcester Polytechnic Institute in Applied Statistics and a B.S. in Mathematics.

LinkedIn:

Twitter: @DataMovesHer

Joy Buolamwini

Algorithmic Bias Researcher | Poet of Code |

Joy Buolamwini uses art and research to illuminate the social implications of artificial intelligence. She founded the Algorithmic Justice League to create a world with more equitable and accountable technology. Fortune Magazine named her to their 2019 list of the world's most outstanding leaders, describing her as "the conscience of the A.I. Revolution" She serves on the Global Tech Panel convened by the vice president of the European Commission to advise world leaders and technology executives on ways to reduce the harms of A.A. technology. In late 2018 in partnership with the Georgetown Law Center on Privacy and Technology, Joy launched the Safe Face Pledge, the first agreement of its kind that prohibits the killer application of facial analysis and recognition technology. She holds two master's degrees from Oxford University and MIT; and a bachelor's degree in Computer Science from Georgia Institute of Technology.

LinkedIn: https://www.linkedin.com/in/buolamwini/

Chip Huyen

Machine Learning Engineer and Open Source Lead at Snorkel AI
Chip Huyen is a computer scientist and writer at a startup that

focuses on the machine learning production pipeline in Silicon Valley.

She helped launch Coc Coc—Vietnam's second most popular web browser with 20+ million monthly active users.

LinkedIn: https://www.linkedin.com/in/chiphuyen/

Twitter: @chipro

Follow her on Medium: Chip Huyen

Fay Cobb Payton

Program Director at National Science Foundation; (Full Tenured) Professor of I.T./Analytics; ACM Education Advisory Board

Fay researches health care, UX design, the bias in computing/I.T. participation, data management and analytics, social & digital inclusion, and others. Her specialties include Health Informatics, Bias in Tech, Social Media, Program Evaluation, Curricula Development, Health Disparities, STEM/ STEM (+Arts) & Workforce Development. She works with I.T. industry professionals interested in retaining, sustaining, and mentoring people of color and executive leadership roles. She has visited 30 U.S. higher education institutions and several in Ghana to shadow executive leadership team members. She was an American Council on Education Fellow, which involved my participation in an academic college review team.

LinkedIn: https://www.linkedin.com/in/cobbpayton/

Kate Crawford

Co-founder of A.I. Now Institute, Senior Principal Researcher MSR, and Distinguished Research Professor at NYU

Kate Crawford is a professor and leading researcher studying data systems' social implications, machine learning, and artificial intelligence.

Also, she collaborates at the École Normale Supérieure in Paris, where she is the inaugural Visiting Chair for A.I. and Justice. At New York University, she is the co-founder of the A.I. Now Institute.

She has advised policymakers in the White House, the Federal Trade Commission, the United Nations, and the City of New York. She co-founded the Fairness, Accountability, Transparency, and Ethics (FATE) group at Microsoft.

As a writer, she collaborates with The New York Times, The Atlantic, The Wall Street Journal, and Harper's Magazine.

She also is a member of the World Economic Forum's A.I. and Robotics Future Council.

Twitter: KateCrawford

Dr. Renata Afi Rawlings-Goss

Executive Director of the South Big Data Innovation Hub | Author | Data Career Coach

Renata Rawlings-Goss is a biophysicist by training. She is the founding Executive Director of the South Big Data Innovation Hub, whose mission is to catalyze partnerships among universities, industry, and government around Big Data, Data Science, and the "Internet of Things" She was awarded the first cohort of AAAS Big Data Policy Fellows supporting congress and the federal government. She was instrumental in creating the National Data Science Organizers group (NDSO.io) and implementing the NSF's priority goal to increase the U.S. workforce in data science. She also serves as the President/ CEO of Good with Data, LLC, which runs The Data Career Academy. She pursues efforts to increase the participation of women and under-represented minorities in STEAM (Science, Technology, Engineering, Art/Design, and Math)

LinkedIn: https://www.linkedin.com/in/afiimani/

Conclusion

As an Artificial Intelligence evangelist and a very data-oriented person myself, I will be thrilled if my daughter will decide to pursue a career in data science, information technology, engineering, or something that might help minimize the gender gap.

I hope the interactions we have and the software experiments we do while we're together will positively affect how she, now and in the future, perceives and uses technology.

I think it's also crucial for young girls like my daughters to be exposed to positive examples of women in Tech, showing them technology as something they can be users and designers or builders. I point out these women to my daughter in our everyday lives, and these awesome women are some examples in this post.

Read more about it...

Here you have a list of other prominent resources for female support in data science and computing below.

Women in Data Science (WiDS) Conference aims to inspire and educate data scientists worldwide, regardless of gender. WiDS started as a one-day technical conference at Stanford in November 2015. Five years later, the WiDS is a global movement that includes several initiatives.

Harvard's Women in Computer Science (WiCS) Advocacy Council—A very active community of faculty and students at Harvard University aiming to understand and reduce the technology gender gap.

Girls Who Code—GWC is a non-profit organization that inspires high-school girls to pursue opportunities in the computing field.

National Center for Women & Information Technology (NCWIT)—A non-profit community of universities, companies, non-profits, and government organizations that aims to increase women's engagement in computing science and technology.

Women in Big Data Forum—This is an active LinkedIn forum to promote diversity in the big data industry by mentoring and peer participation.

Progressive Women's Leadership—A resource center that aims to foster women's leadership in the workplace.

By Jair Ribeiro on September 28, 2020.

22 - Technology vector created by stories - www.freepik.com

What is Reinforcement Learning, and nine examples of what you can do with it?

As part of my new series Short Stories, I will explore some interesting topics in an overview mode, starting from Reinforcement…

Reinforcement Learning is a subset of machine learning. It enables an agent to learn through the consequences of actions in a specific environment. It can be used to teach a robot new tricks, for example.

Reinforcement learning is a behavioral learning model where the algorithm provides data analysis feedback, directing the user to the best result.

It differs from other forms of supervised learning because the sample data set does not train the machine. Instead, it learns by trial and error. Therefore, a series of right decisions would strengthen the method as it better solves the problem.

Reinforced learning is similar to what we humans have when we are children. We all went through the learning reinforcement—when you started crawling and tried to get up, you fell over and over, but your parents were there to lift you and teach you.

It is teaching based on experience, in which the machine must deal with what went wrong before and look for the right approach.

Although we don't describe the reward policy—that is, the game rules—we don't give the model any tips or advice on how to solve the game. It is up to the model to figure out how to execute the task to optimize the reward, beginning with random testing and sophisticated tactics.

By exploiting research power and multiple attempts, reinforcement learning is the most successful way to indicate computer imagination. Unlike humans, artificial intelligence will gain knowledge from thousands of side games. At the same time, a reinforcement learning algorithm runs on robust computer infrastructure.

An example of reinforced learning is the recommendation on Youtube, for example. After watching a video, the platform will show you similar titles that you believe you will like. However, suppose you start watching the recommendation and do not finish it. In that case, the machine understands that the recommendation would not be a good one and will try another approach next time.

Reinforcement Learning Challenges

Reinforcement learning's key challenge is to plan the simulation environment, which relies heavily on the task to be performed. When trained in Chess, Go, or Atari games, the simulation environment preparation is relatively easy. Building a model capable of driving an autonomous car is key to creating a realistic prototype before letting the car ride the street. The model must decide how to break or prevent a collision in a safe environment. Transferring the model from the training setting to the real world becomes problematic.

Scaling and modifying the agent's neural network is another problem. There is no way to connect with the network except by incentives and penalties. This may lead to disastrous forgetfulness. Gaining new information causes some of the ancient knowledge to be removed from the network. In other words, we must keep learning in the agent's "memory."

Another difficulty is reaching a great location—that is, the agent executes the mission as it is, but not in the ideal or required manner. A "hopper" jumping like a kangaroo instead of doing what is expected of him is a perfect example. Finally, some agents can maximize the prize without completing their mission.

Applications areas of Reinforcement Learning

Games

RL is so well known today because it is the conventional algorithm used to solve different games and sometimes achieve superhuman performance.

The most famous must be AlphaGo and AlphaGo Zero. AlphaGo, trained with countless human games, has achieved superhuman performance using the Monte Carlo tree value research and value network (MCTS) in its policy network. However, the researchers tried a purer approach to RL—training it from scratch. The researchers left the new agent, AlphaGo Zero, to play alone and finally defeat AlphaGo 100–0.

Personalized Recommendations

The work of news recommendations has always faced several challenges, including the dynamics of rapidly changing news, users who tire easily, and the Click Rate that cannot reflect the user retention rate. Guanjie et al. applied RL to the news recommendation system in a document entitled "DRN: A Deep Reinforcement Learning Framework for News Recommendation" to tackle problems.

In practice, they built four categories of resources, namely: A) user resources, B) context resources such as environment state resources, C) user news resources, and D) news resources such as action resources. The four resources were inserted into the Deep Q-

Network (DQN) to calculate the Q value. A news list was chosen to recommend based on the Q value, and the user's click on the news was part of the reward the RL agent received.

The authors also employed other techniques to solve other challenging problems, including memory repetition, survival models, Dueling Bandit Gradient Descent, and so on.

Resource Management in Computer Clusters

Designing algorithms to allocate limited resources to different tasks is challenging and requires human-generated heuristics.

The article "Resource management with deep reinforcement learning" explains how to use RL to automatically learn how to allocate and schedule computer resources for jobs on hold to minimize the average job (task) slowdown.

The state-space was formulated as the current resource allocation and the resource profile of jobs. For the action space, they used a trick to allow the agent to choose more than one action at each stage of time. The reward was the sum of (-1 / job duration) across all jobs in the system. Then they combined the REINFORCE algorithm and the baseline value to calculate the policy gradients and find the best policy parameters that provide the probability distribution of the actions to minimize the objective.

Traffic Light Control

In the article "Multi-agent system based on reinforcement learning to control network traffic signals," the researchers tried to design a traffic light controller to solve the congestion problem. Tested only in a simulated environment, their methods showed results superior to traditional methods and shed light on multi-agent RL's possible uses in traffic systems design.

Five agents were placed in the five intersections traffic network, with an RL agent at the central intersection to control traffic signaling. The state was defined as an eight-dimensional vector, with each element representing the relative traffic flow of each lane.

Eight options were available to the agent, each representing a combination of phases. The reward function was defined as a reduction in delay compared to the previous step. The authors used DQN to learn the Q value of {state, action} pairs.

Robotics

There is an incredible job in the application of RL in robotics. We recommend reading this paper with the result of RL research in robotics. In this other work, the researchers trained a robot to learn policies to map raw video images to the robot's actions. The RGB images were fed into a CNN, and the outputs were the engine torques. The RL component was policy research guided to generate training data from its state distribution.

Web Systems Configuration

There are more than 100 configurable parameters in a Web System. The process of adjusting the parameters requires a qualified operator and several tracking and error tests.

The article "A learning approach by reinforcing the self-configuration of the online Web system" showed the first attempt in the domain on how to autonomously reconfigure parameters in multi-layered web systems in dynamic VM-based environments.

The reconfiguration process can be formulated as a finite MDP. The state-space was the system configuration; the action space was {increase, decrease, maintain} for each parameter. The reward was defined as the difference between the intended response time and the measured response time. The authors used the Q-learning algorithm to perform the task.

Although the authors used some other technique, such as policy initialization, to remedy the large state space and the computational complexity of the problem, instead of the potential combinations of RL and neural network, it is believed that the pioneering work prepared the way for future research in this area...

Chemistry

RL can also be applied to optimize chemical reactions. Researchers have shown that their model has outdone a state-of-the-art algorithm and generalized it to different underlying mechanisms in the article "Optimizing chemical reactions with deep reinforcement learning."

Combined with LSTM to model the policy function, agent RL optimized the chemical reaction with the Markov decision process (MDP) characterized by {S, A, P, R}, where S was the set of experimental conditions (such as temperature, pH, etc.), A was the set of all possible actions that can change the experimental conditions, P was the probability of transition from the current condition of the experiment to the next condition and R was the reward that is a function of the state.

The application is excellent for demonstrating how RL can reduce time and trial and error work in a relatively stable environment.

Auctions and Advertising

Researchers at Alibaba Group published the article "Real-time auctions with multi-agent reinforcement learning in display advertising." They stated that their cluster-based distributed multi-agent solution (DCMAB) has achieved promising results and, therefore, plans to test the Taobao platform's life.

Generally speaking, the Taobao ad platform is a place for marketers to bid to show ads to customers. This can be a problem for many agents because traders bid against each other, and their actions are interrelated. In the article, merchants and customers were grouped into different groups to reduce computational complexity. The agents' state-space indicated the agents' cost-revenue status. The action space was the (continuous) bid, and the reward was the customer cluster's revenue.

Deep Learning

More and more attempts to combine RL and other deep learning architectures can be seen recently and have shown impressive results.

One of RL's most influential jobs is Deepmind's pioneering work to combine CNN with RL. In doing so, the agent can "see" the environment through high-dimensional sensors and then learn to interact with it.

RL and RNN are other combinations used by people to try new ideas. RNN is a type of neural network that has "memories." When combined with RL, RNN offers agents the ability to memorize things. For example, they combined LSTM with RL to create a deep recurring Q network (DRQN) for playing Atari 2600 games. They also used RNN and RL to solve problems in optimizing chemical reactions.

Deepmind showed how to use generative models and RL to generate programs. In the model, the adversely trained agent used the signal as a reward for improving actions, rather than propagating gradients to the entry space as in GAN training. Incredible, isn't it?

Conclusion: When should you use RL?

Reinforcement is done with rewards according to the decisions made; it is possible to learn continuously from interactions with the environment at all times. With each correct action, we will have positive rewards and penalties for incorrect decisions. This learning type can help optimize processes, simulations, monitoring, maintenance, and autonomous systems control in the industry.

Some criteria can be used in deciding where to use reinforcement learning:

- When you want to do some simulations given the complexity, or even the level of danger, of a given process.
- To increase the number of human analysts and domain experts on a given problem. This type of approach can imitate human reasoning instead of learning the best possible strategy.

- When you have a good reward definition for the learning algorithm, you can calibrate correctly with each interaction to have more positive than negative rewards.
- When you have little data about a particular problem.

In addition to industry, reinforcement learning is used in various fields such as education, health, finance, image, and text recognition.

Resources

here you have some relevant resources which will help you to understand better this topic:

1. Markov Decision Processes (MDPs)—Structuring a Reinforcement Learning Problem
2. RL Course by David Silver—Lecture 2: Markov Decision Process
3. Reinforcement Learning Demystified: Markov Decision Processes (Part 1)
4. Reinforcement Learning Demystified: Markov Decision Processes (Part 2)
5. What is reinforcement learning? The complete guide
6. Reinforcement learning
7. Applications of Reinforcement Learning in Real World
8. Practical Recommendations for Gradient-Based Training of Deep Architectures
9. Gradient-Based Learning Applied to Document Recognition
10. Neural Networks & The Backpropagation Algorithm, Explained
11. A recurrent neural network-based language model
12. The Elements of Statistical Learning: Data Mining, Inference, and Prediction, Second Edition
13. Gradient Descent For Machine Learning
14. Pattern Recognition and Machine Learning

By Jair Ribeiro on October 23, 2020.

23 - School vector created by vectorjuice - www.freepik.com

What is Lobe, and how is Microsoft Trying to Make AI mainstream?

Microsoft released a free public preview of a tool that lets people train AI models without writing a single line of code.

AI and Machine Learning are complex. We must admit it. And they require advanced knowledge and experience, but today Microsoft started to change this scenario with a very user-friendly tool called Lobe, a free software framework that allows anyone to create machine learning models—no technical skills needed.

Cloud computing, general-purpose GPUs, increased availability of large data sets, and advances in deep learning, a subset of AI machine learning, have sparked a modern AI gold rush. However, the technology's complexity remains an entry barrier for many.

The idea behind Lobe is not new. It started in August 2016 by Mike Matas, Adam Menges, and Markus Beissinger. In September 2018, Microsoft acquired AI startup Lobe to allow anyone to create artificial intelligence.

What is Lobe?

Lobe is a Windows or Mac desktop software program that allows everyone to create machine-learning models for image classification.

It lets you build machine learning models with the help of a simple drag-and-drop interface.

The steps are simple—create a dataset using a web camera or existing images, mark the categories, train the model, evaluate outcomes, then run the model.

Once the model is developed, you can export it to several platforms. Lobe models can be exported as TensorFlow 1.15 SavedModel, a standard format used in Python applications running TensorFlow 1.x or hosted on AWS, Google Cloud, and Azure.

Lobe supports Apple iOS to build iOS, iPad, and Mac apps through Core ML. Exports to TensorFlow Lite support Android / Raspberry Pi smartphone and IoT applications. Lobe supports local, spreadsheet, and photos, and it offers Python and .NET APIs for exports.

What can you do with Lobe?

Microsoft says early users used Lobe to create apps that recognize dangerous plants, detect beehive invaders like wasps, or send warnings when they mistakenly left their garage door open.

Lobe's classification models can be used in many ways. Examples include teaching the software to differentiate toxic from non-toxic plants, responding with emojis to facial expressions, and validating mask wear.

Currently, Lobe supports only image classification projects. Still, I can imagine that object detection, and data classification models can be released in the future.

Conclusion

With Lobe's release to the general public today, Microsoft has taken the first step in taking machine learning to the masses—a change that will accelerate AI from B2B to mainstream customers.

Now, everyone has access to machine learning skills and can create models without needing any technical knowledge.

Could it represent a new step towards artificial intelligence democratization?

Read more about it...

If you want to read more about Lobe, Artificial Intelligence, and Data Science, here you have some other articles I've written about it:

- Microsoft just added 3 interesting new Features to Lobe.
  *Microsoft just released a new version of its tool that lets you train AI models without writing a single line of code.*jairribeiro.medium.com
- 23 Amazing Youtube Channels for you to Learn AI, Machine Learning, and Data Science for Free...
  *This is the perfect moment to start learning something new, and why not start with AI?*medium.com
- Google Objectron—A giant leap for the 3D object detection
  *Google has just announced the launch of MediaPipe Objectron, its mobile technology for real-time detection of 3D...*towardsdatascience.com

By Jair Ribeiro on October 27, 2020.

24 Social media vector created by stories - www.freepik.com

The most impressive YouTube Channels for you to Learn AI, Machine Learning, and Data Science.

This is the perfect moment to start learning something new, and why not start with AI?

I know the pandemic is keeping everyone at home, home working is becoming the new normal for many of us. It is hard to find good presential training these days, but it does not mean that you need to stop learning!

I would say that this is the perfect moment to start learning something new, and why not start with Data Science?

Data Science is a great area to develop your skills today. It combines statistics, Mathematics, Artificial Intelligence (Machine Learning), and data.

You will analyze data, find patterns, make predictions, and help companies solve business problems.

Data Science is a multidisciplinary area focused on data analysis and Machine Learning. This work can feed a web application, for example, but the Data Scientist's job is the analysis and predictive modeling.

The Future of Jobs Reports 2020 edition shows industry-wide parallels while looking at increasingly strategic and increasingly redundant jobs.

Roles, including Data Analysts and Data Scientists, AI and Machine Learning Specialists, Robotics Engineers, and Digital Transformation Specialists, are the leaders in the growing demand for the next future. These are an expanding field of knowledge that has been playing an enormous role in society.

I have selected the best AI, Machine Learning, and Data Science channels on YouTube that I've been using a lot during the last years.

You can still use this crazy 2020 to learn Artificial Intelligence, Python Programming, Machine Learning, Artificial Intelligence, and data science.

SpringBoard

This channel publishes interviews with data scientists from big companies like Google, Uber, Airbnb, etc. From these videos, you can get an idea of what it is like to be a data scientist and acquire valuable advice to apply in your life.

Springboard
*Learn online with a job guarantee. Get a job or your money back. Break into data science, UX design, and more...*www.youtube.com

Arxiv Insights

Xander Steenbrugge is a machine learning researcher at ML6. His YouTube channel summarizes the critical points about machine learning, reinforcement learning, and AI in general from a technical perspective while making them accessible for a bigger audience.

Arxiv Insights
*My name is Xander Steenbrugge, and I read a ton of papers on Machine Learning and AI. But papers can be a bit dry &...*www.youtube.com

Machine Learning 101

A new ML Youtube channel that everyone should check out, Machine Learning 101 posts explainer videos on beginner AI concepts. The channel also posts podcasts with expert data scientists and professionals working on AI in commercial industries.

Machine Learning 101
*This channel was made for two main purposes: to help beginners learn about data science and to help machine learning...*www.youtube.com.

FreeCodeCamp

FreeCodeCamp is an incredible non-profit organization. It is an open-source community that offers a collection of resources that helps people learn to code for free and create their projects. Its website is entirely free for anyone to learn about coding. Also, they have their news platform that shares articles on programming and projects.

freeCodeCamp.org
*Learn to code for free.*www.youtube.com

Data School

Kevin Markham creates in-depth YouTube tutorials to understand AI and machine learning. Data School focuses on the topics you need to master first and offers in-depth tutorials that you can understand regardless of your educational background.

Data School
*Are you trying to learn data science so that you can get your first data science job? You're probably confused about...*www.youtube.com.

Machine Learning TV

Machine Learning TV has resources for computer science students and enthusiasts to understand machine learning better.

Machine Learning TV
*This channel is all about machine learning (ML). It contains all the useful resources which help ML lovers and computer...*www.youtube.com

Giant Neural Network

This YouTube channel aims to make machine learning and reinforcement learning more approachable for everyone. There is a 12 video playlist for a full-introduction to neural networks for beginners. It seems a subsequent intermediate neural network series is currently in production.

giant_neural_network
*Contact: giantneuralnet@gmail.com Discord: https://discord.gg/akUgSGj Making machine learning and reinforcement...*www.youtube.com

Andreas Kretz

Andreas Kretz is a data engineer and founder of Plumbers of Data Science. He broadcasts live tutorials on his channel on how to get hands-on experience in data engineering and videos with questions and answers about data engineering with Hadoop, Kafka, Spark, and so on.

Andreas Kretz
I help you get into data engineering, the plumbing of data science. Building up big data platforms. Home of the...www.youtube.com

Edureka!

Edureka is an e-learning platform with several tutorials and guidelines on trending topics in the areas of Big Data & Hadoop, DevOps, Blockchain, Artificial Intelligence, Angular, Data Science, Apache Spark, Python, Selenium, Tableau, Android, PMP certification, AWS Architect, Digital Marketing and many more.

edureka!
Thank you for Subscribing! If you have not, Subscribe now! We are a live & interactive e-learning platform with the...www.youtube.com

Andrew Ng

Ng was named one of Time's 100 Most Influential People in 2012 and Fast Company's Most Creed. He co-founded Coursera and deeplearning.ai and was a former vice president and chief scientist at Baidu. He is an adjunct professor at Stanford University.

Andrew Ng
Enjoy the videos and music you love, upload original content, and share it all with friends, family, and the world on...www.youtube.com

Deeplearning.ai

The official Deep Learning AI YouTube channel has video tutorials from the deep learning specialization on Coursera. Founded by Andrew Ng, DeepLearning.AI is an education technology company that develops a global AI talent community.

DeepLearning.AI's expert-led educational experiences provide AI practitioners and non-technical professionals with the necessary tools to go all the way from foundational basics to advanced application, empowering them to build an AI-powered future.

Deeplearning.ai
*Welcome to the official deeplearning.ai Youtube channel! Here you can find the videos from our Deep Learning...*www.youtube.com

Tech with Tim

Tech With Tim is a brilliant programmer who teaches Python, game development with Pygame, Java, and Machine Learning. He creates high-level coding tutorials in Python.

Tech With Tim
*Python Programming, Game Development, Pygame, Java Tutorials, and Machine Learning. This is a list of a few of the...*www.youtube.com

Machine Learning University (MLU)

Created in 2016, Machine Learning University (MLU) is an initiative by Amazon with a direct objective: to train as many employees as possible to master the technology, essential for the company to achieve the "magic" of offering products with this integrated technology.

Machine Learning University
*Welcome to the channel for Machine Learning University! Our mission is to make machine learning accessible to anyone...*www.youtube.com

Artificial Intelligence—All in One

This YouTube channel has tutorial videos related to science, technology, and artificial intelligence.

Artificial Intelligence - All in One
*All video tutorials related to science and technology will be updated here. LIKE, SUBSCRIBE, SHARE to distribute it...*www.youtube.com

Sentdex

Sentdex creates one of the best Python programming tutorials on YouTube. His tutorials range from beginners to more advanced. With more than 1000 videos about Python Programming tutorials, it goes further than just the basics. You can learn about machine learning, finance, data analysis, robotics, web development, game development, and more.

sentdex
*Python Programming tutorials, going further than just the basics. Learn about machine learning, finance, data analysis...*www.youtube.com

Joma Tech

Joma Tech is a YouTuber who makes videos to help people get into the technology industry. He worked for large technology companies as a data scientist and software engineer. Based on his experience, he makes videos of interviews with experts and lifestyle in Silicon Valley and makes data science more accessible.

Joma Tech
*I talk about life in Silicon Valley, big tech companies, data science, and software engineering. Joma Startup: A web...*www.youtube.com

Python Programmer

Python Programmer content includes tutorials on Python, Data Science, Machine Learning, book recommendations, and more.

Python Programmer
*Hi, I'm Giles McMullen-Klein, and this is my YouTube channel. My degree was in physics, which is my main interest, but...*www.youtube.com

Deep Learning TV

This YouTube channel features topics such as how-to's, reviews of software libraries and applications, and interviews with key individuals in the field of deep learning. DeepLearning.TV is all about Deep Learning, the field of study that teaches machines to perceive the world. Starting with a series that simplifies Deep Learning, the channel features topics such as How To's, reviews of software libraries and applications, and interviews with key individuals in the field. Through a series of concept videos showcasing every Deep Learning method's intuition, we will show you that Deep Learning is simpler than you think.

DeepLearning.TV
*DeepLearning.TV is all about Deep Learning, the field of study that teaches machines to perceive the world. Starting...*www.youtube.com

Google Cloud Platform

YouTube videos to help you build what's next with secure infrastructure, developer tools, APIs, data analytics, and machine learning, Helping you build what's next with secure infrastructure, developer tools, APIs, data analytics, and machine learning.

Google Cloud Platform
Helping you build what's next with secure infrastructure,

*developer tools, APIs, data analytics, and machine learning...*www.youtube.com

Keith Galli

Keith Galli is a recent graduate from MIT. He makes educational videos about computer science, programming, board games, and more.

Keith Galli
*Recent MIT Graduate. I make educational videos on Computer Science, Programming, Board Games, and more! I found online...*www.youtube.com

Data Science Dojo

Data Science Dojo is a channel that promises to teach data science to everyone in an easy to understand way. You will find a multitude of tutorials, lectures, and courses on data engineering and science.

Data Science Dojo
*At Data Science Dojo, we believe data science is for everyone. Our in-person data science Bootcamp has been attended by...*www.youtube.com

Updates from our readers

TechnoBotic

This is a very nice Youtube Channel for Machine Learning, Data Science, and Artificial Intelligence. The videos are about AI and ML basics as well as Advanced Concepts. In this channel, The author summarizes his core learnings from a practical and usability perspective while making them accessible for a larger audience. If you love Machine Learning, Data Science, and Artificial Intelligence, this channel is for you.

StatQuest

StatQuest breaks down complicated Statistics and Machine Learning methods into small, bite-sized pieces that are easy to understand. StatQuest doesn't dumb down the material. Instead, it builds you up to have a better understanding of Statistics and Machine Learning.

Statistics, Machine Learning, and Data Science can sometimes seem like very scary topics, but since each technique is…www.youtube.com

Yannic Kilcher

Yannic makes videos about machine learning research papers, programming, and issues of the AI community and the broader impact of AI in society.

I make videos about machine learning research papers, programming, and issues of the AI community and the broader…www.youtube.com

Conclusion

These channels are unique. I've been following all of them for a long time. I'm always fascinated by the tremendous amount of knowledge we can get today online for free.

I hope you enjoy it. If you know any other engaging YouTube channels about AI, Machine Learning, Deep Learning, or Data Science, leave it in the comments!

Read more about it…

If you want to go further on your learning journey, I've prepared for you a great list with more than 60 training courses about AI, Machine Learning, Deep Learning, and Data Science that you can do right now for free:

- The best free courses to learn AI, ML, and Data Science today.
  More than 60 courses with ratings and a summary (Made by AI, of course).jairribeiro.medium.com
- Is it the end of the work as we know it?
  *A brief analysis of the report Future of Job 2020 by the World Economic Forum*medium.com

By Jair Ribeiro on November 2, 2020.

25 - Business vector created by pch.vector - www.freepik.com

Europe leads the way on set rules for Artificial Intelligence

A framework to balance between protecting citizens and fostering technological development.

The EU has released recommendations on artificial intelligence regulation to balance protecting customers and to promote technological growth. These include an agreement on IP problems, a development ethics policy, and liability rules setting penalties of up to € 2 million and a 30-year limitation period for such claims.

These are amongst the first detailed legislative proposals to be published internationally, making for interesting reading for stakeholders worldwide. The recommendations cover three areas:

- an ethics framework for AI
- liability for AI causing damage
- intellectual property rights

For AI product producers, these ideas merit careful consideration. Specifically, those running "high-risk" AI face the possibility of a rigorous new regulatory regime. Next year, the European Commission said it would issue draft regulations on AI.

The Commission could well adopt any of the European Parliament's proposals, or variants on them. Affected stakeholders will have opportunities to engage with any new AI laws during the normal legislative process. Still, efforts to understand how these proposals could affect your company should start now.

An ethics framework for AI

The legislative initiative by Iban García del Blanco (S&D, ES) urges the EU Commission to present a new legal framework outlining the ethical principles and legal obligations to be followed when developing, deploying, and using artificial intelligence, robotics, and related technologies in the EU including software, algorithms, and data.

Future laws should be made following several guiding principles, including a human-centric and human-made AI, safety, transparency, and accountability; safeguards against bias and discrimination; right to redress; social and environmental responsibility; and respect for privacy data protection.

High-risk AI technologies, such as those with self-learning capacities, should be designed to allow for human oversight at any time.

Suppose functionality is used to result in a severe breach of ethical principles and could be dangerous. In that case, the self-learning capacities should be disabled, and full human control should be restored.

Liability for AI causing damage

Axel Voss (EPP, DE) 's legislative initiative calls for a future-oriented civil liability framework, making those operating high-risk AI strictly liable for any resulting damage. A clear legal framework would stimulate innovation by providing businesses with legal certainty while protecting citizens and promoting their trust in AI technologies by deterring activities that might be dangerous.

The rules should apply to physical or virtual AI activity that harms or damages life, health, physical integrity, property, or that causes significant immaterial harm if it results in "verifiable economic loss." While high-risk AI technologies are still rare, MEPs believe that their operators should hold insurance similar to that used for motor vehicles.

Intellectual property rights

The report by Stéphane Séjourné (Renew Europe, FR) makes clear that EU global leadership in AI requires a significant intellectual property rights system (IPR) and safeguards for the EU's patent system to protect innovative developers while stressing that this should not come at the expense of human creators' interests, nor the European Union's ethical principles.

MEPs believe it is essential to distinguish between AI-assisted human creations and AI-generated creations. They specify that AI should not have a legal personality; thus, ownership of IPRs should only be granted to humans. The text looks further into copyright, data collection, trade secrets, the use of algorithms, and deep fakes.

Conclusion

With this framework, the European Parliament is among the first institutions to put forward recommendations on what AI rules should include ethics, liability, and intellectual property rights.

Future laws should be made following several guiding principles, including a human-centric and human-made AI.

By Jair Ribeiro on November 3, 2020.

26- Blue vector created by vectorjuice - www.freepik.com

A gentle Introduction to Data Literacy

Data literacy is a relatively recent trend that simultaneously in business that making informed decisions. Let's find out more about it.

With the evolution of Artificial Intelligence and other technological innovations, many companies neglect what remains the main asset of any business: its human capital. Humans generate data; data is the new oil; it is the new currency.

If you have the impression that you hear it before… The phrase "data is the new oil" is not mine.. but it was said—and has been repeated since it was coined by the British mathematician Clive Humby in 2006—to denote this value and power in the data in our business lives.

Data is not just numbers; they are texts, videos, images, audios, and all kinds of information encoded. Data are even people. The combination of millions, billions, trillions of information make up what we call Big Data.

What is Data Literacy

Data literacy is a relatively new concept that emerged simultaneously in companies using business intelligence to make better decisions.

It's the idea that everyone should know how to make decisions with data to perform their function to the fullest.

Gartner defines data literacy as "the ability to read, write and communicate data in context, including an understanding of data sources and constructs, analytical methods and techniques applied—and the ability to describe the use case, application and resulting value."

Data literacy is a perfect analogy to a person's literacy process in their mother tongue: it requires effort, repetition, reading, accompanying professionals in the field, interest ... But, once this step is taken, the process flows. And when people are literate, they read, write, work, communicate, think, reason, and argue in that language. Data literacy works the same way.

The data has no meaning in itself. Like a letter or a number, you can take the data out of context and make it almost meaningless.

A routine but straightforward example: a Google search with only the word "Plant." The results are disconnected and varied, mentioning: planting soybeans, planting a banana tree, planting a chip, how to plant a tree ... So, it is necessary to contextualize and specify the searches so that the data found are more assertive and accurate.

Data—*without the interpretation of people*—has no meaning in itself. It is the combination of human reasoning ability with artificial intelligence that generates some conclusions about the information. Although data analysis can provide a significant differential for the organization in several aspects, one must not lose sight of what is behind all this, precisely the people.

Data scientist and analyst Susan Etlinger speak in her TED talk: "We have the opportunity to give the data meaning by ourselves.

Frankly, data does not create these senses, we do". ***And how can we create these meanings for the data?***

> *"We are not passive consumers of data and technology. We shape the role they play in our lives and the way we give meaning to them. But to do that, we need to pay as much attention to how we think about the way we code. We have to ask questions tough questions, to move from the moment of 'telling' things to the moment of understanding them", according to data scientist* Susan Etlinger.

The Data Literacy Project

Recently, Qlik and Accenture surveyed more than 9,000 interviews (with people from different industry segments, from C-light to beginners, nine countries in North America, Europe, and the Asia Pacific) and produced a report about the current scenario of data literacy in companies.

The report brings a somewhat worrying reality: although managers consider that they are developing actions to be guided by data (data-driven), the breakneck pace of some changes is eventually causing feelings of anxiety, fear, overload, uncertainty, and even sadness among employees.

According to the survey data, 74% of respondents reported feeling unhappy (unprepared, insecure) when working with data.

The machines can process an increasing volume of data and information with more sophistication and agility, making thousands of combinations and even imitating human beings' behavior, suggesting specific responses to a service's users.

Thus, it is inevitable that they will be increasingly desired and used to automate various companies' processes. It turns out that, as mentioned above, this has caused distress and overload for employees, who have to deal with a mountain of information, updates, and new tools every day.

Even if they feel uncomfortable when dealing with data, they recognize that it is an asset to the organizations they work for and believe that data literacy training would make them more productive.

Data literacy allows operations leaders to communicate success measures to the c-suite. It helps finance convey urgency when the sales function is not meeting quarterly and yearly targets.

Improving business data literacy positively affects gross margin, return on assets, return on equity, and return on sales, resulting in a $ 320 to $ 534 million difference in companies' values that focus and do not focus on Data Literacy (data provided by The Data Literacy Index).

The study also found that 76% of top business decision-makers do not trust their Data Literacy skills.

The impact of Data Literacy

Therefore, the critical point is to include people, involve all sectors in data analysis, break barriers, and overcome technical issues to democratize and demystify data access. In this way, the company can benefit as a whole.

This also impacts relevant aspects of culture, stimulating the exchange of experiences and knowledge, a collaboration between peers, regardless of the area, and continuous learning.

Having a healthy and robust culture and preparing people to deal smoothly and efficiently with data-based decisions are strategic issues to build a more humane and pleasant corporate environment—characteristics increasingly valued by the market—and optimize results, including financial.

Dissatisfied and anxious employees may unconsciously end up contaminating the company's culture and becoming demotivated.

When accounted for the slowness in the execution of medical processes and licenses resulting from stresses resulting from issues associated with data and technology, companies lose an average of more than five working days (43 hours) per employee per year.

Of course, changes in culture do not happen overnight, but it is a fact that it is in constant motion. Knowing this and how much this is a fundamental pillar, managers must always be attentive, seeking to share this responsibility to develop a fair culture based on values that make sense to the teams, in line with the organization's interests.

Thus, digital inclusion becomes a less stressful process and with better results, since people will be more open and willing to incorporate new tools into their work routine, understanding more deeply not only the functionalities related to these technologies and data, but the gains that she can have with this and, consequently, being more genuinely interested in analysis, data, etc.

Therefore, before investing millions in data usage projects, invest in people, as they are the key to data science strategies and other projects. If it is necessary to take a step back or slow down the pace of implementation of the analyzes to educate employees about data more cautiously, including making it attractive to these people— and not only to those used to the data -, no hesitate. The results in the medium and long term will be justified.

Conclusion

As we can see, Data literacy is the idea that everyone should know how to make decisions with data to perform their function to the fullest.

It has a relevant impact on aspects of our business culture, stimulating the exchange of experiences and knowledge, the collaboration between peers, regardless of the area, and continuous learning.

Companies lose an average of more than five working days (43 hours) per employee per year due to stress-related data and technology. The only way to reduce this stress is with data education and awareness.

Data awareness as a second language is a necessity of the present, not just of the future. Are you and your company working to make it happen? Tell me about it in the comments!

By Jair Ribeiro on November 6, 2020.

27 - Infographic vector created by katemangostar - www.freepik.com

How AI and Digital Transformation will change your business forever.

Artificial Intelligence means for the digital transformation what electricity has meant to humanity in the past. Are you ready?

Digital Transformation is one of the most critical drivers on how companies will continue to deliver value to their customers in a highly competitive and ever-changing business environment.

Artificial Intelligence (AI) has been recognized as one of the central enablers of digital transformation in several industries.

The transformation process seeks to leverage digital technologies to create or modify customer experiences and culture, and business processes, thus meeting customers' changing needs and the market.

And this is where AI comes into play. It can help companies become more innovative, more flexible, and more adaptive than ever.

The promise of speed, ease, and cost optimization, while simplifying complex processes and systems, places artificial intelligence as one of the most significant digital transformation drivers.

And although many consider it as a technology of the future, it is already here, being used by many companies looking to optimize their business.

So, let's see how Artificial Intelligence can help your business as one of the most potent enablers of what we call Digital Transformation. But what is Digital Transformation?

Defining the Digital Transformation scenario

Digital transformation is a set of processes, methodologies, and tools used by modern companies to optimize their operational activities, such as providing differentiated service, increasing performance, and increasing its reach power, with employees and customers as a priority.

However, digital transformation is not just a new department in the organization. Still, it is a game-changer in technology's role in the corporate environment. That's why it is currently being considered as the 4th Industrial Revolution.

But more than a concept, digital transformation has become a movement that attracts companies interested in reviewing processes, innovating, and gaining competitiveness with the help of technology.

In the context of transformation, technology is not an end. Still, it is a set of tools that need to be at the service of the company's business strategy.

And today, no matter in what industry your business is operating, with a considerable probability, your business use technology to deliver products or services.

With a very similar probability, your competitor is technology-based too, and they can come from any segment.

But on the other side, there is still a lot of technology investment to be done, and the impact hasn't even started.

Data and Artificial Intelligence (AI) are critical factors in the strategy for those who want to expand their business impact in this digital transformation journey. Data only makes sense if it is aligned with the process and seen as the company's competitive advantage.

Before talking about AI, do you speak data?

Getting value from data is at the heart of any digital business transformation.

The enormous volume of data, coming from different sources and formats, such as structured (ERP, Database, etc.) and unstructured (social media) data, if treated correctly, can help your business to understand better the desires of your customers, the market where they operate and your competitors, bringing insights for increasingly intelligent and agile decision making.

The use of data is a central point in the management of companies in their digital transformation. It is essential to have a strategy to use them as a means of having a competitive advantage.

For example, using more advanced analysis, based on qualified data, to beat the competition.

Building an efficient and comprehensive Data Literacy is the only way to be effective on the real scale and bring insights to the business in a collective effort to make sense of all that information.

It is necessary to establish processes and resources capable of connecting, structuring, and analyzing this data.

Companies understand the value of investments in innovative technologies and processes capable of analyzing data more efficiently and quickly—allowing them to reap these investments' benefits.

With the possibility of integrating different systems and automating several daily tasks, the digital transformation took another leap when Artificial Intelligence (AI) and Machine Learning (ML) became part of many organizations' business strategies.

In addition to resulting in faster and more efficient operations and, therefore, more productivity, these technologies are so important in the digital transformation because they allow better use of the data collected by your company in several ways.

In a reality in which 90% of all data produced in history has been generated in the past two years, it is necessary to make sense of them. As the famous saying goes, "data is the new oil."

Machine Learning and AI allow us to use all this amount of information to take the company further, either by improving current products and services or by the possibility of new innovative strategies.

Undoubtedly, the most significant impact is the learning that the machines gave to the human being, a much more excellent notion about the scenario that we are inserted in.

Artificial Intelligence (AI) and Machine Learning (ML) are two of the most potent digital transformation protagonists. They are the basis for the most efficient digital tools developed today. They are enablers of increasingly innovative and effective solutions that directly impact the market's acceleration and competitiveness and customers' experience and expectations.

Where to start?

A persistent question that may come to your mind regarding data is what you should do with it and where you should start.

I will try to add my two cents to help you answer those, but also, you may be in the condition to ask other common questions like:

- How can data improve my customers' experience?

- Should I hire someone?
- Should I invest in a database?
- How do I know I have enough data to generate intelligence?

I should say that the first job to be done, which is also a challenge, is to gather all your data in an organized way and start processing it so that it becomes strategic information for your team or company.

From experience, I believe that there is no lack of data. But it is true the opposite—the contrary. "There are many sources of internal and external data, but everything is spread out, different platforms, databases, silos, paper piles, everything is loose across your company. You need to find it and organize it.

Defining your journey to Artificial Intelligence

Following the path opened by digital transformation, there is no way to have artificial intelligence without having a clear and defined data strategy.

There is no point in seriously talking about Artificial Intelligence if you do not have your data organized.

It would be best if you did your homework before starting experiments with AI and Machine Learning.

Start importing the data and storing it efficiently, preferably on the cloud. Today is becoming more challenging to keep the crescent amount of data produced by any business organized sustainably on a server or in a data center. Cloud is the direction you will probably want to follow.

Investing in Artificial Intelligence is to train the machine and the algorithms from databases that are organized for that to happen. To help this task, an ever-growing computational power is available on the cloud to solve anything.

The following is a proposal of the necessary steps to reach the stage of maturity in the use of AI to optimize its production process with all the potential available effectively, based on the SingleStore Maturity framework:

1. Collect data: first of all, it is necessary to have data; AI is hugely dependent on data, examples, and instruments that serve as samples for training models that assist in decision making. This data can be extracted from databases, spreadsheets, markup files such as XML, etc.
2. Storage: in addition to collecting data from safe and quality sources, it is necessary to use tools for storage, structuring, and integration that facilitate data exploration analysis. Here are ETL-type tools (Extract, Transform, and Load) responsible for extracting, transforming, and even loading data. Such devices are essential to prepare the data for the next stage of exploration.
3. Exploration: in this step, descriptive analyzes are made. BI (Business Intelligence) reports, datamart tools, OLAP queries (Online Analytical Processing), and analysis panels are built from the data collected and stored so that specialists can have a clearer, more compact, and objective view of sectors and the operation of the organization as a whole.
4. Real-time operation and extraction: this is a level of maturity at which the organization is concerned with integrating its data with modern tools, many of them on servers in the cloud, using APIs (Application Programming Interface, the transformation of data in formats of easier integration like JSON (JavaScript Object Notation) or XML. Such strategies facilitate integration in real-time and improve the response time from creating data to more improved analysis.
5. Prediction and Optimization: at this stage of maturity, the organization already has quality data, in real-time, in formats compatible with the leading technologies used for training machine learning models and is capable of making decisions based on analysis and prediction. High-level algorithms are developed at this level, capable of recognizing voice, images, recommending, learning from patterns, etc.

How can you create value with Digitization through Artificial Intelligence?

Just as the advent of the internet has changed the way the world has always done business, emerging technologies, especially artificial intelligence, are gradually entering the daily lives of businesses worldwide.

The question is no longer when the digital transformation will arrive in all small and medium-sized companies, but what technologies will be essential and prioritized.

Artificial intelligence can be used in corporations in various industries. The following examples can serve as a basis for technology companies' leaders to better understand opportunities to provide SME services.

To create value with a product, digital transformation must be applied, with artificial intelligence identifying information to help the involved professionals in various stages of a process, such as design, execution, and delivery.

With everything digitalized, you may start checking the phases that need adjustments in the project; for example, in the design phase, AI improves research, development and makes an accurate forecast of the next steps.

In the execution phase, continuous maintenance is the critical point where AI and Machine Learning can help. These are responses in real-time to what is being researched.

When delivering a product, it is, as previously mentioned, the customer experience that you will provide with everything digital, AI, and Machine Learning can be used to monitor, recommend and forecast actions to reinforce your product, brand, and market share.

The importance of an early adoption

McKinsey estimates AI techniques can create between $3.5T and $5.8T in value annually across nine business functions in 19 industries, generating up to $2.6T additional value in Marketing and Sales and up to $2T in Supply Chain Management and Manufacturing.

Also, AI will add $200B in value to Pricing & Promotion and $100B to Customer Service Management in Retail.

McKinsey predicts AI will have an 11.6% impact on Travel industry revenues and up to 10.2% on High Tech.

And in most of these use cases, AI and deep neural networks improved performance beyond what existing analytic techniques were able to deliver.

As you can see, there are several reasons for you and your business to bet quickly on Artificial Intelligence technologies, and I like to focus on three of the most relevant:

1. The necessity to monitor and combat legacy's "technological DNA," which guarantees agility to innovate and deliver services quickly;

2. The necessity to offer products and services increasingly directed to the individual needs of customers, in an omnipresent and assertive manner;

3. The potential to multiply the business value through automated processes and offers increases operations efficiency.

It is also important to mention that, by strengthening these operational capacities with Artificial Intelligence, your business will also enhance its resilience to global crisis scenarios by offering products and services with increased profitability and creating predictive capabilities for their business.

Conclusion

Artificial Intelligence means to the digital transformation what electricity has meant to humanity in the past.

Its disruptive power is so great that we move towards a stage in the economy where digital products will be increasingly intelligent to make recommendations, present options, and help customers make their choices.

The biggest challenge for all of us is to manage all these changes and deal with such a transformation in the organizational structure.

Investing and developing skills among the entire workforce to adapt to new models and trends is critical to bring positive results.

Are you ready for the revolution?

By Jair Ribeiro on November 11, 2020.

28- Car vector created by vectorjuice - www.freepik.com

Honda will bring the level three autonomous vehicles to the masses.

Honda will be first to mass-produce level 3 autonomous cars by the end of March 2021

Honda claims it will be the world's first automaker to mass-produce sensor-packed level 3 autonomous cars that will allow drivers to let their vehicles navigate congested expressway traffic, meeting the SAE Level 3 standards.

The automaker has plans to produce and sell a version of its Honda Legend luxury sedan with fully approved automated driving equipment in Japan from next March.

They announced the news via press release (via Reuters), and this follows the approval by the Japanese government of the company's "Traffic Jam Pilot" autonomous tech, which for the first time, will allow drivers to take their eyes off the road while it's engaged.

The race to build self-driving cars is a crucial technology battleground for automakers. Technology companies such as

Google parent Alphabet Inc also invest billions of dollars in a field expected to boost car sales.

Autonomy Levels

Autonomous driving is a technology that will change how we will travel and move goods across the world.

Reduced traffic congestion, lower travel costs, and no more circling for parking spaces will make our daily commutes quicker, less stressful, and more affordable. It'll also reduce harmful CO_2 emissions, improving the quality of air that we breathe.

Different cars are capable of varying levels of autonomy, described on a scale of 0 to 5, and essential to an understanding before we talk about an autonomous vehicle's operation.

The more technological solutions in actuators and sensors the automobile incorporates, the greater its degree of automation. As there are several stages in development, regulations and technical definitions also need to adapt.

For this reason, the Society of Automobile Engineers (SAE) created a classification to differentiate vehicles according to their degree of automation, making it easier for consumers and maintenance professionals to identify the models. The following five levels have been determined:

Level 0: *Your car today*—humans control all significant systems.

Level 1: *Driver assistance*—specific systems, such as cruise control or automatic braking, can be controlled by the vehicle, one at a time. At this level, the driver still handles most of the car's functions but with a little autonomous help. For example, a level one vehicle might provide you with a brake boost if you edge too close to another vehicle, or it might have an adaptive cruise control function to control your distance and speed.

Level 2: *Partial automation*—the vehicle offers at least two simultaneous automatic functions, such as acceleration and steering, but requires human beings for safe operation. Partial automation enables drivers to disengage from some driving functions. Level 2 vehicles can assist with tasks like steering, acceleration, braking, and maintaining speed. However, drivers still need to have both hands on the wheel and be ready to take control if necessary.

Level 3: *Conditional automation*—the vehicle can manage all critical safety functions under certain conditions. Still, the driver must take over when alerted. At this level, cars can be considered truly autonomous, but only under ideal road conditions.

Level 4: *High automation*—the vehicle is fully autonomous in some driving scenarios, although not all. At Level 4, vehicles are capable of steering, accelerating, and braking on their own. They're also able to monitor road conditions, respond to obstacles, determine when to turn, and change lanes.

Level 5: *Full automation*—the vehicle is fully capable of autonomy in all situations, requiring no human interaction. Vehicles can steer, accelerate, brake, and monitor road conditions like traffic jams. Essentially, Level 5 automation enables the driver to sit back and relax without paying any attention to the car's functions whatsoever.

Read more about it...

If you want to read more about Autonomous Vehicles, please have a look at these articles:

- An Introduction to Autonomous Vehicles
  *Autonomous vehicles have long lived in our imagination since the Jetsons, and if we can imagine, we can do it. The...*jairribeiro.medium.com
- How Autonomous Vehicles will redefine the concept of mobility.
  Autonomous cars are already among us, and some actions

*have already been taking regarding auto repair shops and dealer...*towardsdatascience.com
- **The Ethics of AI and Autonomous Vehicles**
  *In a perfect world, AI should be developed to avoid unethical issues, but that may be unlikely since those issues can...*medium.com

By Jair Ribeiro on November 12, 2020.

29- Business vector created by katemangostar - www.freepik.com

Five companies that are revolutionizing recruiting using Artificial Intelligence

Virtual recruiters and advanced analytics tools come with a proposal for a safer and more transparent hiring process.

Artificial intelligence (AI), the use of human-like intelligence through software and mechanisms, enables the disruption of diverse segments. After all, this is an industry that has grown an average of 20% per year for the past five years, according to a survey by BBC Research.

Many organizations have already joined the "future" and gained space by efficiently applying AI in everyday activities. For example, some banks started to perform financial services without a human's help; farms use drones to identify points in a crop that need more irrigation and automatically trigger sprinklers.

AI is not set to replace the recruiter's work, the importance of the interview, the empathy, and the sparkle in the eye that we sometimes feel when interviewing a candidate. This area deals with human relations, and this will hardly be replaced.

However, AI applications arrive to make life easier for recruiters and allow HR to play an increasingly strategic role.

Considering that HR departments are called to be in line with young people's expectations, digital natives are used to more intuitive and technological experiences.

Can you imagine having access to a virtual assistant that assists your enrollment in selection processes? Today, this is becoming a reality! Chatbots, gamification, and Artificial Intelligence applications are ways of no return for companies that always want to be ahead of Talent Acquisition competitors.

Artificial Intelligence can also be used to test candidates' skills and behaviors in real situations, generating an incredibly smarter database for making hiring decisions, reducing their acquisition cost, and making the Human Resources area much more strategic and less operational.

Some applications of artificial intelligence in recruitment and selection

The insertion of technology in the recruitment and selection processes has several applications and brings excellent facilities to those involved.

For example, the specific and slow steps of analyzing curricula one by one and separating the best ones. With AI, this stage ceases to exist in the whole field and is now entirely done in the virtual world.

Specialized software automatically evaluates curricula, selecting those that best meet the requirements of the vacancies.

AI can also automate the online application of skills tests, determining the most appropriate questions to get to know each candidate more deeply.

Applications of competency tests and behavioral profiles can also be made through AI. In this way, there is a guarantee that only the most suitable professionals for a given position reach the end of the process, the personal interview.

Five disruptive companies that are revolutionizing recruiting using AI

Here you have five great examples of companies that are using AI to break the ground on recruitment today:

HireVue

HireVue is arguably the best-known AI-powered hiring platform, deployed by 700+ companies, including Unilever, Vodafone, PwC, and Oracle. For example, Unilever deployed HireVue AI-driven assessments and achieved £1M annual cost savings, a 90% reduction in time to hire, and a 16% increase in hiring diversity.

Along with voice and facial recognition software, HireVue has a proprietary algorithm to determine which candidates are ideal for a specific job by analyzing their vocabulary, speech patterns, body language, tone, and facial expressions.

The sophisticated machine learning algorithms are implemented by a strong team of data scientists headed by Dr. Lindsey Zuloaga.

Mya Systems

Hiring chatbot Mya Systems uses conversational AI to streamline the recruiting process for staffing agencies and companies such as L'Oréal, Adecco, Hays, and Deloitte.

Mya guides candidates through the entire hiring process, starting from the job search and up to the onboarding. Mya leverages state-of-the-art approaches from natural language processing and understanding to allow natural conversation with the candidates, supported by a team of experienced machine learning engineers and NLP engineers.

HiredScore

HiredScore transforms how companies hire and retain employees by providing AI-driven solutions that seamlessly integrate with the clients' existing HR systems, ensuring compliance and security.

HiredScore uses machine learning to understand how large companies hire candidates and develop unique insights to provide grades to new applying candidates and let the company's recruiters focus on the candidates that match their jobs the best. Also, HiredScore circulates high-quality leads that might have been rejected in past processes or signed up to receive potential job offers from the company in the future.

The company's proprietary AI is customized for each client, proactively mitigates bias, and is trained on a large dataset that includes 25M CVs, 50M job posts, and 21K career trees. Deep learning techniques are developed and implemented by an experienced team of applied ML researchers and data scientists.

Wade & Wendy

Wade & Wendy's AI-driven solutions automate task-driven recruitment processes for both job seekers and recruiters. [1]

Wade & Wendy's intelligence is driven by a proprietary recruiting conversation system, cutting-edge text parsing techniques, intelligent workflow automation, and a robust knowledge graph consisting of conversational utterances, linguistic logic, and job seekers' attributes, candidates, and job positions.

To achieve the best performance for their conversational system, NLP data scientists at Wade & Wendy experiment with cutting-edge models, including BERT and XLNet.

Hiretual

Hiretual offers yet another set of AI-powered comprehensive solutions for most recruiting activities.

Hiretual can be easily integrated with 30+ Applicant Tracking Systems (ATSs) to save time with one seamless workflow, sync candidate activities, and manage duplicates. The solution is supported by smart business analytics and industry-standard security and compliance measures.

Conclusion

The fact that AI is advancing and occupying more and more space in our daily lives is nothing new; it is present from the time our cell phone rings up in the morning to wake us up and permeates the rest of our day, whether at home, on the street, or at work.

The challenges faced by HR in the contemporary world require constant readaptations. The old management models will not be able to meet the demands that arise. For this reason, managers are increasingly talking about disruptive thinking.

One of the most significant advantages of AI is eliminating the factor "personal opinion of the recruiter" (or guesswork, guesswork, and prejudice) in hiring.

Another benefit is the ability to deal with many resumes objectively and efficiently. But not only that.

Virtual recruiters and advanced analytics tools come with a proposal for a safer and more transparent selection process, breaking the ties with inconsistent and laborious methods and moving forward to modern and digitalized HR Services.

References:

1. *7 AI Companies Revolutionizing Recruiting, by Kate Koidan—TopBot*
2. *Why I've decided to join HiredScore by Dennis Nerush—HackerNoon*

By Jair Ribeiro on November 14, 2020.

30- Work vector created by stories - www.freepik.com

Twelve (+Bonus) amazing YouTube Channels to Learn Python Programming for Free

Here you have a great list of Youtube channels to dig into Python programming and learn from the best.

A few days ago, I published an article with 21 of the best channels on Youtube where you can learn Data Science, AI, and Machine Learning for free and... Boom! It was a great success; many people wrote to me that the article was handy and helped them find great content!

Suppose you are looking for the best Youtube channels to dig into Python programming and learn from the best. In that case, you have a great list with 12 (my lucky number) amazing programmers who share tips and secrets that will help you become a master!

Clever Programmer

You can find awesome programming lessons here! Also, expect programming tips and tricks that will take your coding skills to the next level.

Clever Programmer
*You can find awesome programming lessons here! Also, expect programming tips and tricks that will take your coding...*www.youtube.com

Anaconda Inc.

Source: screenshot from the related Youtube Channel

With over 4.5 million users, Anaconda is the world's most popular Python data science platform. Anaconda, Inc. continues to lead open source projects like Anaconda, NumPy, and SciPy, forming modern data science. Anaconda's flagship product, Anaconda Enterprise, allows organizations to secure, govern, scale, and extend Anaconda to deliver actionable insights that drive businesses and industries forward."

Anaconda, Inc.
*With more than 15 million users, Anaconda is the world's most popular data science platform and the foundation of...*www.youtube.com

Talk Python

Talk Python to Me is a weekly podcast hosted by Michael Kennedy. The show covers a wide array of Python topics as well as many related issues.

Talk Python
*Talk Python to Me is a weekly podcast hosted by Michael Kennedy. The show covers a wide array of Python topics as well...*www.youtube.com

Christian Thompson

Christian Thompson and a lot about Python programming for beginners: Christian is a middle and high school teacher who uses

Python as his teaching language. He firmly believes anyone can (and should) learn to program a computer and that Python is the perfect language for doing so.

TokyoEdTech
*Welcome to my channel! My channel is dedicated to teaching coding in a way that is beginner-friendly, fun, and...*www.youtube.com

CodingEntrepreneurs

Coding for Entrepreneurs is a Programming Series for Non-Technical Founders. Learn Django, Python, APIs, Accepting Payments, Stripe, jQuery, Twitter Bootstrap, and much more.

CodingEntrepreneurs
*Coding for Entrepreneurs is a Programming Series for Non-Technical Founders. Learn Django, Python, APIs, Accepting...*www.youtube.com

Corey Schafer

Source: screenshot from the related Youtube Channel

This channel is focused on creating tutorials and walkthroughs for software developers, programmers, and engineers. We cover topics for all different skill levels, so whether you are a beginner or have many years of experience, this channel will have something for you.

Corey Schafer
*Welcome to my Channel. This channel is focused on creating tutorials and walkthroughs for software developers...*www.youtube.com

Chris Hawkes

On this channel, you can learn about programming, web design, responsive web design, Reactjs, Django, Python, web scraping, games, forms applications, and more!

Chris Hawkes
*We're going to learn about programming, web design, responsive web design, Reactjs, Django, Python, web scraping...*www.youtube.com

Enthought

For more than 15 years, Enthought has built AI solutions with science and engineering at the core. We accelerate digital transformation by enabling companies and their people to leverage the benefits of Artificial Intelligence and Machine Learning."

Additionally, Enthought is best known for the early development, maintenance, and continued support of SciPy and the primary sponsor for the SciPy US and EuroSciPy Conferences.

Enthought
*For more than 15 years, Enthought has built AI solutions with science and engineering at the core. We accelerate...*www.youtube.com

Real Python

Python tutorials and training videos for Pythonistas that go beyond the basics. On this channel, you'll get new Python videos and screencasts every week. They're bite-sized, and to the point, so you can fit them in with your day and pick up new Python skills on the side.

Real Python
*Python tutorials and training videos for Pythonistas that go beyond the basics. 🎓 Get free Python tips and...*www.youtube.com

Sentdex (Harrison Kinsley)

Python Programming tutorials, going further than just the basics. Learn about machine learning, finance, data analysis, robotics, web development, game development, and more.

sentdex

*Python Programming tutorials, going further than just the basics. Learn about machine learning, finance, data analysis...*www.youtube.com

Python Basics | Learn Python Programming

Learn the basics of python programming language. Simple and easy to learn.

Python Basics

*Learn the basics of python programming language. Simple and easy to learn.*www.youtube.com

Telusko

The channel was started in the year 2014 and now teaches various Programming topics. The video lectures include Core Java, Advanced Java, Python, Android Development, Blockchain, JavaScript, and many other languages.

The Python playlist has more than 100 videos, which will take you from the basics to the advanced language. You can also learn about the Django and Flask technologies to advance Python Learning.

Al Sweigart

Al Sweigart

*Enjoy the videos and music you love, upload original content, and share it all with friends, family, and the world on...*www.youtube.com

PythonBytes

PythonBytes

*Python bytes is all about the Python programming language, building GUI's and web framework applications.*www.youtube.com

By Jair Ribeiro on November 19, 2020.

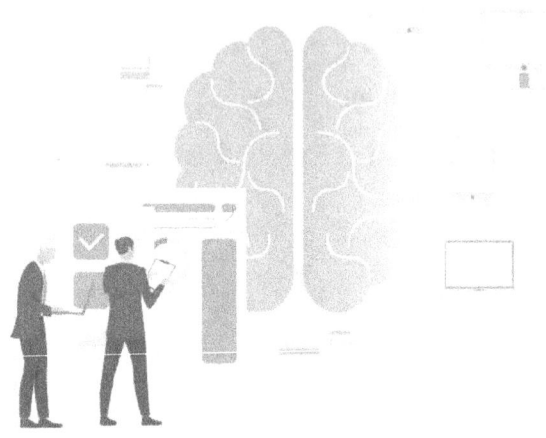

31 - Infographic vector created by katemangostar - www.freepik.com

A Simple Approach to Define Human and Artificial Intelligence

Before you start worrying about AI, you should consider analyzing the relationships between human intelligence and AI.

I recently started to follow an exciting and mind-bending philosophy online course at MIT called Minds and Machines.

The course is a thorough, rigorous 12 Weeks Learning Path introduction to contemporary philosophy of mind, exploring consciousness, reality, artificial intelligence (AI), and more. It is definitively one of the most in-depth philosophy courses available online that I ever frequented.

The first effect of starting study philosophy at Massachusetts Institute of Technology is that I'm asking more challenging questions… the second effect is that I'm writing more about those questions.

I'm in this moment, exploring the relationship between the mind and the body, the capacity of computers to think, the way we perceive reality, and the perspective of the existence of a science of consciousness.

As a first result, I've started to pay particular attention to one specific question that definitively has a lot to relate to my daily work as an AI expert: what is intelligence?

In this article, I will explore human and artificial intelligence concepts to find relevant similarities and differences.

A long time question…

The idea of building devices that could simulate the movements of living beings and particularly of humans in a completely autonomous way dates back to the Upper Paleolithic.

The making of primitive dolls with movable arms was one of the first attempts to imitate living beings' gestures. Archimedes was a master in this sort of new "art."

Fantastic literature, known today as science fiction, speculated that artificial sentient beings could be made.

Robotics has made science fiction a full-fledged science known as "artificial intelligence" or robotics in recent decades.

The purpose of these researches is to create a sentient being endowed with decision-making abilities and move at will—a sentient being and to a certain extent independent of human beings., but the real question here is…

What is Intelligence?

The term intelligence comes from the Latin intelligentĭa, which, in turn, derives from *inteligere*. This is a word that is composed of two other terms: *intus* ("between") and *Legere* ("choose"). Therefore, the etymological origin of the concept of intelligence refers to those who know how to choose: intelligence allows them to select/choose the best options for solving a question.

Intelligence is a set that forms all the intellectual characteristics of an individual, that is, the faculty of knowing, understanding, reasoning, thinking, and interpreting. Intelligence is one of the main distinctions between humans and other animals.

Etymologically, the word "intelligence" originated from the Latin *intelligentia*, creating from *intelligere*, in which the prefix *inter* means "between," and *Legere* means "choice." Therefore, this term's original meaning refers to an individual's ability to choose among the various possibilities or options presented to him.

To choose the best and most appropriate opportunity, among the various options, a person needs to evaluate to the maximum all the advantages and disadvantages of the hypotheses, requiring this the ability to reason, think and understand, that is, the basis of what forms intelligence.

Among the faculties that constitute intelligence, there is also the functioning and use of memory, judgment, abstraction, imagination, and conception.

The concepts and definitions of intelligence vary according to the group to which they refer. For example, in psychology, the so-called "psychological intelligence" is the ability to learn and relate, that is, an individual's cognition; while in the field of biology, "biological intelligence" would be the ability to adapt to new habitats or situations.

Types of intelligence

However, the Intelligence Quotient concept started to be discredited when individuals with low IQ were observed, but with great professional life success. At the same time, people considered "more intelligent" presented terrible situations.

The psychologist Howard Gardner presented the Theory of Multiple Intelligences, which claims that intelligence is a set of at least eight different mental processes existing within the brain.

According to this theory, each human being has a little bit of each one of these "bits of intelligence." In some people, there is always a specific type of process that can be more developed than in others, making it stand out in particular fields or activity areas.

- Linguistic intelligence: people who can easily express themselves, orally, and through writing. People with this more developed type of intelligence tend to learn other languages more easily, in addition to having a high degree of attention.
- Logical intelligence: people with ease in working with logic in general, such as mathematical operations or scientific works. They usually have a good memory and can solve complex problems quickly. They can also be considered more organized and disciplined.
- Spatial intelligence: people with ease in understanding and manipulating the visual world, such as 2D or 3D images. Architects and graphic art professionals develop them well.
- Motor intelligence: people who can perform complicated movements with their bodies have a fantastic notion of space, distance, and depth of environments.
- Musical intelligence: people who quickly identify and reproduce different types of sound patterns, in addition to creating new songs or harmonies. This is one of the rare kinds of intelligence present among people.
- Interpersonal intelligence: people who are easy to lead, based on understanding the point of view and others' intentions. They are considered very active individuals who enjoy responsibilities and find it easy to convince others to do what they want.
- Intrapersonal intelligence: people who can observe, analyze, and understand themselves. They can also influence people, but more subjectively, using ideas and not actions.
- Naturalistic intelligence: people who can quickly identify and differentiate different patterns present in nature.

Emotional intelligence

The concept of emotional intelligence is present within psychology and was created by the American psychologist Daniel Goleman.

An emotionally intelligent individual can identify his emotions, motivating himself to persist in his goals even in frustrating situations.

Among the other characteristics of emotional intelligence are controlling impulses, channel emotions into appropriate situations, motivating people, and practicing gratitude, among other qualities that can encourage others.

Artificial intelligence

Intelligence relates to knowing how to choose the best options to solve some problem. According to their attributes and processes, several intelligence types are biological intelligence, operational intelligence, and psychological intelligence.

AI is also an adjective that is said of what is made by hand, art, or the ingenuity of man. The artificial also makes it possible to refer to what is unnatural or false.

Artificial intelligence was developed about specific systems created by human beings that constitute non-living rational agents. In this case, rationality is understood as the ability to maximize an expected result.

Therefore, artificial intelligence consists of designing designs that produce results that maximize performance when executed on a physical architecture.

These are processes based on specific sequences of entries that are perceived and stored by said architecture.

Devices with artificial intelligence can perform various processes analogous to human behavior, such as executing a response for each input, searching for a state among all possible ones according to action, or resolving problems through formal logic.

"AI—Artificial intelligence" is also a film made by Steven Spielberg, whose debut took place in 2001. His argument is based on the story of a robot that, when created to replace a human child, demonstrates having feelings.

Artificial intelligence or AI is a computer science branch of study that deals with developing technological mechanisms and devices that can simulate human beings' reasoning system, i.e., intelligence.

Research related to artificial intelligence is slow, but it has shown significant results on how devices can interpret and synthesize human voice or movements. There is still a long way to go before machines reach the concept as close as possible to human intelligence.

Human Intelligence vs. Artificial Intelligence

There are several differences between artificial and human intelligence, ranging from cognitive to emotional and psychological issues.

Talking about the differences between artificial intelligence (AI) and humans can be reduced because the former was created by the latter. However, there are more differences that we are going to tell you so that you will understand more how it is almost impossible for AI to surpass the natural intelligence of the human being that delves into cognitive functions such as memory, problem-solving, learning, planning, language, reasoning, and perception.

Although today both play an enormous role in improving societies, there are apparent differences.

AI is an innovation created by human intelligence and is designed to perform specific tasks much faster with less effort.

Human intelligence is best at multitasking and can incorporate emotional elements, social interaction, and self-awareness into the cognitive process. The latter is characterized by being highly complex such as concept formation, understanding, decision making, communication, and problem-solving. It is also heavily influenced by subjective factors like motivation.

Human intelligence or HI is commonly measured through IQ tests that generally cover working memory, verbal comprehension, processing speed, and perceptual reasoning.

When compared to humans, computers can process a lot more information at a faster rate. For example, if the human mind can solve a math problem in five minutes, AI can solve ten problems in one minute.

> *"The view that machines will think as man does reveal a misunderstanding of the nature of human thought."* Ulrik Neisser.

The AI is very objective in decision-making since it analyzes using data collected purely. However, human decisions can be influenced by subjective elements that are not based only on figures.

Another difference may be that artificial intelligence often produces accurate results, as they operate based on a set of programmed rules. When it comes to human intelligence, there is generally room for "human error," as specific details can be lost at one point or another.

Human intelligence is, in general, very flexible in response to changes in their environment. This enables people to learn and master various skills. On the other hand, AI takes much longer to adapt to new changes.

The human intellect can support multitasking, as evidenced by diverse and simultaneous roles. In contrast, AI can only perform a few tasks simultaneously, as a system can only learn responsibilities one at a time.

Artificial intelligence is still working on self-awareness, while humans naturally become aware of themselves and strive to establish their identities as they mature.

As human beings, we are much better at interacting with other people, considering that we can process abstract information, have self-awareness, and be empathetic. On its way, Artificial Intelligence has not mastered the ability to pick up relevant social and emotional cues.

Human intelligence's general function is innovation, as it can create, collaborate, generate ideas, and implement. As for AI, its overall role is more of optimization. It performs tasks efficiently according to the way it is scheduled. These are the main differences between machines and humans.

Conclusion

The idea of intelligent machines capable of reaching human intelligence levels is relatively recent compared to how long humans have been able to think. However, the advanced techniques applied not only to computer science but also to biology and engineering, among others, have already become fundamental to our society.

But modeling human intelligence from computer systems becomes a challenging goal to create sensitive and evolvable machines. When establishing relationships between human intelligence and artificial intelligence, we must start from the hypothesis that there is a computational model of which we have more information related to its operation and memory capacity, and this may lead us to consider the possibility, in a general way, of whether artificial intelligence will surpass human intelligence in the future.

Human intelligence's general function is innovation, as it can create, collaborate, generate ideas, and implement. As for AI, its overall role is more of optimization. These could be considered the main differences between machines and humans.

References:

- Description Of The Word Of Intelligence—900 Words | Bartleby. https://www.bartleby.com/essay/Description-Of-The-Word-Of-Intelligence-PKQSQDKVG5YW
- What is the Meaning of Artificial intelligence https://edukalife.blogspot.com/2015/07/what-is-meaning-of-artificial.html
- ARTIFICIAL INTELLIGENCE V/S HUMAN INTELLIGENCE!—Google-Wizz. https://googlewizz.com/artificial-intelligence-v-s-human-intelligence/
- What Kind Of Smart Are You? Question 5—You're.... https://www.quizony.com/what-kind-of-smart-are-you/5.html
- Difference Between Artificial Intelligence & Human http://www.authorstream.com/Presentation/FuGenx2008-4014030-difference-artificial-intelligence-human/
- Difference Between Artificial Intelligence and Human .... http://www.differencebetween.net/science/difference-between-artificial-intelligence-and-human-intelligence/
- Howard Gardner, Theory of Multiple Intelligences
- Neisser, Ulric. "The Imitation of Man by Machine." *Science* 139, no. 3551 (1963): 193–97. Accessed November 28, 2020. http://www.jstor.org/stable/1710006.

By Jair Ribeiro on November 28, 2020.

32- Cartoon vector created by vectorjuice - www.freepik.com

What is Predictive Analytics, and how can you use it today?

To see the future, you can rely on two tools: a crystal ball or Predictive Analytics.

Predictive analytics is a way to use the past to project the future of your business. This is not futurology, but an accurate calculation of the probabilities in any scenario, based on the processing of large volumes of data.

This advanced technique uses data mining, machine learning, and artificial intelligence to further statistics. Rather than concluding about yesterday, you can anticipate trends and predict tomorrow's behaviors—all from your company's history.

Want to understand how predictive analytics helps you make informed decisions? After this text, you will know how to add predictive analysis to your business to start ahead of the competition.

What is predictive analytics?

Predictive analysis is an advanced analytical technique that uses data, algorithms, and machine learning to anticipate trends and make business projections. Thanks to computational advancement, it is already possible to analyze large volumes of data (Big Data) to find patterns and evaluate future possibilities from its history.

The concept originated in the 1940s when governments started using the first computers—those that occupied an entire room and served warlike purposes.

But, predictive analytics has gained far more relevance today, driven by powerful processors and new technologies.

Another decisive factor for the rise of this technique is Big Data: the phenomenon of accelerated multiplication of information with 2.5 quintillion bytes of data being produced by humans every day. And if you are asking, there are 18 zeroes in a quintillion. You're welcome!

Therefore, analytics's function is to situate ourselves in this immensity of data, showing the possible directions to follow and looking for patterns in the middle of the whirlwind of information.

The predictive analysis uses data mining, machine learning, artificial intelligence, and statistics to collect, process, interpret and translate it.

But it is essential to clarify that this technology cannot "predict the future," only to map the probabilities based on what has already occurred.

The essential question is not "What will happen?", But "What is likely to happen?".

One of the most basic examples of applying this type of analysis is cross-selling—the strategy of encouraging the customer to add complementary products and services at the time of purchase.

Do you know that famous e-commerce recommendation: "people who bought this product also took..."?

In companies, it is possible to use predictive analysis systems to predict possible customer behaviors based on their purchase history, interactions, and profile.

Thus, product recommendations are much more accurate, thanks to the reliable forecast generated by crossing millions of data.

Similarly, the tool can be used to predict the acceptance of a new product on the market, understand which marketing strategies are most promising, and anticipate operational failures.

How does predictive analytics work?

There are several possible approaches, but, as a rule, the concept is based on creating a predictive model. This mathematical function will predict a problem when applied to the data.

For example, a pharmaceutical laboratory can apply a predictive model on your order history to decide whether to increase the production of a particular drug next winter considering the weather estimates for the period (a stricter, drier, rainier season), anyway).

Similarly, companies can use predictive models to determine whether a particular product has a good chance of success, whether switching suppliers can streamline the production cycle, whether consumers will well receive a change in packaging, and so on.

It is worth noticing that machine learning (or machine learning) can play a crucial role in predictive analysis. Like? Machine learning is a system that modifies its behavior autonomously based on patterns found in data sets. Because of this, it is common for algorithms of this type to be developed or adapted to act specifically in predictive analysis.

Importance of predictive analysis for companies

With increased competitiveness and profound changes in the digital age, companies need—more than ever—to be one step ahead of the competition.

In this case, using predictive analysis is like having a strategic vision of the future, mapping the opportunities and threats that the market has in store.

Therefore, companies are adopting predictive models for:

- Predict the next moves in the segment
- Identify opportunities ahead
- Prevent security breaches
- Optimize marketing strategies
- Map the behavior and habits of consumers and employees
- Improve operations and increase efficiency
- Reduce risks.

You can use predictive analytics to understand a consumer's likely behavior, optimize internal processes, monitor and automate IT infrastructure and machine maintenance, for example.

Not by chance, the global predictive analytics market is forecast to move $ 10.95 billion by 2022, according to a report published in 2018 by Zion Market Research. After all, nothing is better for business today than making decisions based on reliable analysis.

About Predictive Analytics, Big Data, and Business Intelligence.

As crucial as obtaining data is knowing how to use it.

Big data is the primary source of research for the construction of predictive models. The choice of data, or data mining, consists of identifying which records and statistics can build the best strategic information.

On the other hand, business intelligence can be a sector within the organization chart or even its strategy. Its function is to transform or refine the data to transform it into information, which, in a way, allows its name to be used in such a generic way.

Predictive analytics applications need to be fed with lots of data, turning them into useful information and creating continuous improvement processes. There is a mutual exchange between data and analysis; one cannot live without the other.

Data analysts can create predictive models when they have enough data to obtain predicted results. Therefore, all matters are deeply related.

What are predictive models?

We already know that predictive analysis uses data from the past and present to predict future behavior through statistical functions. They are also able to detect patterns in the analyzed data set.

A predictive model is what a predictive modeling professional creates using relevant data and statistical methods. These models can be used to answer specific questions and predict unknown values.

Predictive models fall, in general, in two fields: parametric and nonparametric. While these terms may sound like technical jargon, the main difference is that parametric models make more and more specific assumptions and assumptions.

Some of the types of predictive models are:

- Ordinary Least Squares;
- Generalized linear models;
- Logistic regression;
- Random forests;
- Decision trees;
- Neural networks;
- Multivariate adaptive regression splines.

Each of these models is used for a specific purpose; that is, it answers a particular question or type of data set.

In short, all models have methodological and mathematical differences and are similar in their shared objective, which is the prediction of future or unknown results.

How to Do Predictive Analytics in 7 Steps

To understand how predictive analytics works in practice, let's follow the main steps of the process.

See how to apply the concept in 7 steps.

Definition of objectives

To create a predictive model, you need to start from a project with well-defined business objectives.

To begin, you should ask yourself what the purpose of the analysis is:

- Understand consumer behavior?
- Predict sales trends?
- Identify the most profitable products?
- Reduce churn rate or turnover?
- Reduce production and operating costs?
- Reach a new target audience?

Definition of analysis goals

The next step is to translate your business objectives into analysis goals.

For example, if you want to understand consumer behavior better, you must create a predictive profile analysis model.

Other possible models are risk analysis, segmentation, activation, customer lifetime value (LTV), etc.

Data collection

With the objectives and goals defined, it's time to hunt for the data needed to answer your questions.

This step requires the most care. It is the quality of the data that will define your predictive analysis's reliability.

Therefore, you need to select the best sources to collect the data (internal databases, social networks, research, consultancy databases) and define precisely what information is required.

It is crucial to use an adequate collection tool and determine the data's accuracy, cost, and stability.

Data preparation

Before starting the analysis, it is essential to prepare the data so that they are in the correct format and can be read by your tool.

Start with cleaning up unnecessary information, define variables, sort your data, and then structure them into specific sets.

You can do this with software such as Excel and Power BI, for example.

Data analysis

With the data properly structured, you can now begin the analysis process.

At this point, it is important to have statistical knowledge to evaluate the resulting graphs and understand your trend line.

For example, suppose you analyze customer transaction data. In that case, you will have a clear view of the hottest periods, best selling products, and possible influencing factors on sales variation.

Here, you have three basic analysis options:

- Univariate analysis: each variable is treated in isolation before being crossed with the others
- Bivariate analysis: establishes a relationship between two variables (Ex .: sale time and average ticket)
- Multivariate analysis: establishes relationships between two or more variables (Ex .: customer's age, LTV, and average ticket).

Modeling

After conducting your analysis and carrying out the necessary tests, you are ready to create a predictive model with this data.

This model will be a standard of mathematical and statistical techniques that processes the data collected from the relationships you have created, offering quick and easy to view responses.

Thus, your predictive analysis will begin to return valuable insights into future probabilities.

Monitoring

After creating your predictive model, you should closely monitor its efficiency to ensure that the results remain reliable.

Ideally, the models' performance should be reviewed monthly, quarterly, and semi-annually to ensure that a possible change in data does not affect the analysis.

Three examples of Predictive analysis software

These are examples of great software that can be your great ally when implementing predictive analysis. Have a look at these three that are some of the most used in the market.

Microsoft Power BI

Power BI is a famous data analysis software and Microsoft business intelligence. With it, you can import data directly from Excel spreadsheets or data warehouses and conduct high-performance predictive analysis.

Also, the tool allows the resulting reports and graphs to be shared with everyone on the team.

Adobe Analytics

Adobe Analytics is the strong competitor of Google Analytics, with a unique tool for predictive analytics.

The system uses machine learning and statistical modeling to analyze data in an advanced way and to predict future behaviors such as turnover and conversion probability.

Thus, you can gain insights from various data sets and prepare for future scenarios.

Tableau

Tableau is one of the BI platforms (Business Intelligence) market leader with advanced predictive analytics.

Tableau's key feature is the ability to modify calculations and test different scenarios in sophisticated analyzes, using various sets, groups, and segmentations.

All of this in a simple panel, with drag and drop commands, makes it easy for even inexperienced users to use.

Five practical examples of how predictive analytics can be applied today.

Predictive analytics is a highly versatile method that can be used in many areas of your company.

See what the leading applications are.

Customer Churn forecast

Making a churn forecast means identifying the signals that precede your customers' cancellation request, calculating the probability in each situation.

With predictive models, you can cross-check data such as customer service quality, customer satisfaction level, and churn rate to determine which factors influence cancellation.

The goal is to understand the main reasons for the customer's loss and reverse this process.

Campaign optimization

Your entire history of marketing campaigns can be used to project better results in the future.

Just use predictive analytics to identify the best channels for each content, the most successful language for each target audience, and other variables predicting consumer acceptance.

Thus, you aim directly at the target when engaging and winning over your audience.

Lead segmentation

Predictive analytics is also great for creating lead segmentation strategies.

After all, one of the biggest challenges of marketing is to map the profile of these potential customers to offer tailored content and create infallible nutrition campaigns.

With data and machine learning, you can generate segmented groups based on sophisticated analysis, predicting what leads want to receive the smallest details.

Customer Relationship Management (CRM)

In CRM strategies, you can use predictive models to understand customers' every moment during the life cycle and purchase journey.

In this case, there is no lack of data to create multivariate models and analyze the most diverse possible relationships between behaviors, profiles, purchase history, interactions, and customer perceptions.

With these powerful insights in hand, you can revolutionize your customer relationship with personalized content, promotions, and offers.

Fraud detection

Analytical methods also allow companies to detect patterns of fraud and prevent security breaches. With the discussion of cybersecurity on the rise, more and more organizations are concerned with correcting vulnerabilities and identifying any abnormalities in time to avoid damage.

Predictive models make it much easier to identify threats in real-time and anticipate scams.

Risk management

Risk management is another area that directly benefits from the predictive analysis.

It would be much easier to make decisions with a complete view of the risks and opportunities ahead, right?

Therefore, predicting the probabilities of profit or loss is the great differentiator of advanced data analysis, whether to analyze a customer's credit risk or the possible consequences of an investment.

Conclusion

Did you understand the importance of predictive analysis to see the future of your business?

Of course, the data has no clairvoyant power. Still, it is possible to map the possibilities for making better decisions and going beyond your competitors.

With the impressively fast evolution of AI and machine learning, the tendency is for algorithms to become increasingly intelligent and to make even more accurate predictions.

As we have seen, human intelligence is indispensable in the process, as you need to feed systems with quality data to obtain good results.

Read more about it...

If you want to learn more about AI, Machine Learning, and Data Science, check this article:

- 21 amazing Youtube channels for you to learn AI, Machine Learning, and Data Science for free
  This is the perfect moment to start learning something new, and why not start with AI? towardsdatascience.com

References

1. What is Predictive Analytics?—Enterprise IT Definitions .... https://www.hpe.com/us/en/what-is/predictive-analytics.html
2. How Much Data Is Created Every Day in 2020?.... https://techjury.net/blog/how-much-data-is-created-every-day/
3. Global Predictive Analytics Market expected to reach USD https://www.globenewswire.com/news-release/2017/01/12/905404/0/en/Global-Predictive-Analytics-Market-expected-to-reach-USD-10-95-Billion-by-2022-Zion-Market-Research.html

By Jair Ribeiro on December 4, 2020.

33 - Car vector created by vectorjuice - www.freepik.com

Will My Daughters Ever Drive a Car?

My daughters will grow as Autonomous Passengers in a new kind of mobility, but will they need a driver's license?

I am the father of three amazing daughters: Stella, a beautiful and smart 13 years old teenager, Sofia, an incredibly smart three years old and last but not least… Emily, just one old but already able to show all her strong and curious character.

These girls do their best to keep me busy with all those kinds of stuff that fathers must deal with every day. Maybe it is too early to get worried about it.. but recently, after a pleasant conversation with a friend that works like me in the transportation industry, I've started to hear in my mind an almost philosophical question:

My daughters are digital natives, but they still use a very analogic and traditional piece of paper to draw their amazing and fantastic animals and flowers at home and many paper notebooks while learning at school… so the question is: will they need a driver's license in 2038?

Well, considering how things are developing today, probably not… Let me tell you why…

The dream of a new mobility

Imagine waking up early for work, getting ready, reading some news online on your tablet while having breakfast, without worrying about traffic and the chance of being late for having been in the traffic jam for forty minutes.

You access an application on your cell phone and request a car to take you to work, and in a few minutes, it arrives. A mechanical voice says a polite "good morning, welcome," confirms that the destination is entered in the application. The vehicle starts the journey by driving automatically.

The AI algorithm will calculate the shortest route considering the variables of traffic, traffic lights, number of vehicles, accidents, and works on the way. Whatever else may affect the trip. Comfort will be a priority: air temperature, water if you feel thirsty, and various playlists to choose from will be available.

This scenario seems to be the future of mobility.

My daughters will grow in this reality. Considering that, as they are deemed Native Digital today… they will grow as *Autonomous Native Passengers* in a new kind of mobility, ruled by Autonomous Vehicles.

Safety as a horizon for new mobility.

I love to imagine this autonomous scenario for the future, which may not happen anytime soon, considering that full automation faces "very complex" problems. It could end up reducing these vehicles to some very restricted applications.

But I am a dreamer, and I believe that Autonomous Vehicles' technology is an indispensable tool for zero accidents on roads worldwide. This will be the leading driver (pun intended) for massive adoption.

I was born in Brazil at the end of the '70s. I had the not desirable fortune to grow in a country where traffic incidents used to count more deaths than some military conflicts.

Over the years, I've had lost an unacceptable number of friends due to traffic incidents.

Recent studies show some improvements, but the situation is still dramatic, even with some increase in the number of fatal incidents.

According to the World Health Organization, in 2018, car accidents were the leading cause of young people's death up to 29 years old. To give you an example, in the USA, 94% of accidents, which caused 1.35 million deaths, had their causes related to human errors.

It is argued that, by 2025, autonomous cars will represent 4% of the total vehicles sold in the world and 75% in 2035, within 15 years. This could drastically reduce the number of incidents across the globe. I will tell you why…

Autonomous Vehicles will drive better than you do.

Autonomous vehicles will not get tired, distracted, or intoxicated while driving.

They will be controlled by AI systems connected to redundant sensors and other top-level electronic equipment.

They go from one place to another, as the user instructed. On the way, they will collect all the necessary environmental information, such as signals, pedestrians, and other vehicles, while being directed by satellite systems to make a safe and optimized trip.

This technology has been developed globally, in universities and research centers, and the automotive industry.

Many brilliant minds are putting a tremendous amount of money and effort into making it happen safely and efficiently.

Rethinking our cities

If this technology succeeds, and I really believe it will, not only will my daughters' driving experiences will change, but the whole metropolitan model wherever they will decide to live in the next 20 years will change, too.

According to a study carried out in the United Kingdom, shared VAs will increase the urban space by 15 to 20%, mainly by eliminating parking areas.

Research is being published on the international scene with optimistic projections of the autonomous transport application in different instances.

In the next 20 years, I see that it will be much more comfortable and pleasant to live in cities once they start adapting to Autonomous Vehicles.

This transformation is already taking place. It will allow no accidents on the road in the next 20 years, but this will depend on a process of profound change based on three pillars where shared. Connected mobility, autonomous and electrical will rely on technology in vehicles also outside of them.

My daughters will grow in a complex metropolitan ecosystem formed by drones that take people or feed agriculture, robots, and various vehicles of different sizes that can take many people and just one (the so-called micro-mobility), wireless electrical charges, and other technologies we already start to see today.

But even with all this available automation, will they be able to or required to drive? The question is still open.. and maybe to help us to answer that, we should look through the next point:

What are the advantages of AVs?

As we can see today, the main advantages of Autonomous vehicles are safety enhancements and time savings.

For example, today drive for an hour to get to work could dedicate themselves to work and do video conferences in their vehicles…

Hopefully, my daughters in 2038 will use their commuting time to study, watch some programming language on youtube or chat with their friends and colleagues, thanks to the increased comfort and security delivered by the AVs.

They will probably not have a long commuting time since AVs are expected to reduce traffic and reduce accidents due to the remarkable optimization that AI algorithms will apply to our mobility.

For sure, their health will have a great benefit from less pollution in our cities, because 100% of AVs in the future will be electric.

And what about the driver's license?

But what happens if, instead of all these features and improvements provided by full automation, my daughters will decide to put themselves behind the steering wheel, considering that still, a steering wheel will be available in AVs vehicles?

Well… I don't see this as an alternative in 2038 when Emily, my younger daughter, will be 18 years. According to research carried out in the UK in 2019, drivers of self-driving automobiles will need certification to adapt to new vehicles.

According to the survey, the driver's license will continue to be indispensable in the future since vehicles will require human intervention in certain circumstances.

This kind of study is beneficial to put some light on the limitation of technology today. However, I still believe that in the next 20 years, many of the open questions still represent a roadblock to the full adoption of AVs will find their reliable answers. We will reach a level of standardization and safety requirements in our roads that will be unacceptable to allow humans to drive.

Probably we will move the ownership of the driver's license from humans to the algorithms directly. But this is another story that I will cover in a future article.

Conclusion

Self-driving vehicles will become part of our lives. When controlled by algorithms instead of humans, vehicles will lose their current value as a fetish or status symbol and finally become a tool.

AVs' benefits are numerous and significant. Their adoption will represent a great challenge to many business models as we know today.

Maybe it is too early to imagine VAs on a large scale on the streets, but it is time to put some questions and start to prepare the ground for them in all spheres.

Autonomous Vehicles will be a reality very soon. In 1908, Henry Ford revolutionized the way we see cars, making them a symbol of utilitarianism, comfort, and status. Now, after just over a hundred years, AI will reinvent mobility and start a new era.

Read more about it

If you want, you can read more about Autonomous Vehicles in these articles:

- An Introduction to Autonomous Vehicles
  *Autonomous vehicles have long lived in our imagination since the Jetsons, and if we can imagine, we can do it. The...*towardsdatascience.com
- How Autonomous Vehicles will redefine the concept of mobility.
  *Autonomous cars are already among us and some actions are already been taking regarding auto repair shops and dealer...*towardsdatascience.com
- The Ethics of AI and Autonomous Vehicles
  *In a perfect world, AI should be developed to avoid unethical issues, but that may be unlikely since those issues can...*medium.com

By Jair Ribeiro on December 5, 2020.

34 - People vector created by pikisuperstar - www.freepik.com

Thoughts about the Need for More Diversity and Inclusion in AI

AI has a diversity and inclusion problem, but the good news is that AI itself can give us a hand.

I'm black, and I am a foreigner living in Poland, a country that still has a long road to cover regarding diversity and inclusion of minorities and I've been very active in volunteering on a local ONG, designing and delivering workshops that help hundreds of teenagers better understand the value of tolerance and inclusion in our society.

I am also a father of three young girls. I've been very committed to encouraging girls and women in computing because I'm sure that diversity—not only of gender but racial and social—is fundamental to the area in which I work: Artificial Intelligence.

Because I believe that without the right amount of diversity, AI will not be able to reasonably contribute to solving the most relevant problems and deepen our world's inequality. Maybe you are asking yourself why it matters…

Well… let me tell you simply that AI is already impacting your life every day. Everything you do… so probably you would prefer that these ubiquitous and almighty algorithms should be fair, ethical, and inclusive; just life should be. If you are still not convinced, I will try to give you something to think about…

Why does the lack of diversity in Artificial Intelligence can be an issue?

We can count several cases of prejudice in AI systems, including improper classification of minorities, chatbots who have learned to disseminate hate speech (Microsoft), black people being classified as gorillas in search engines (Google), etc.

An interesting report called Discriminating Systems—Gender, Race, and Power in AI by the University of New York draws attention to facts that deserve discussion: many "flaws" related to cases of discrimination widespread in Artificial Intelligence (AI) systems would be associated with the lack of diversity in the teams that work these technologies.

Unfortunately, this lack of diversity in AI negatively affects systems. Many works with automatic data learning and bias (ideological, gender, race, etc.).

To get an idea of this lack of diversity in AI teams, here are some data from the report:

- More than 80% of AI teachers are men;
- Only 15% of AI researchers on Facebook and 10% of AI researchers at Google are women;
- Men currently represent around 71% of the group of candidates for IA US jobs, as shown in the 2018 AI Index report.

As Rob Doyle wrote in an interesting article about Sexism in Tech: An Inconvenient Truth:

> To tackle sexism in the tech industry, first, the status quo must change. Then, it must be consistently monitored at all levels. This involves solving problems in the current workplace and getting things

> right at an educational level. More females need to be encouraged to learn STEM subjects to create an equilibrium in the tech industry—Rob Doyle

This lack of diversity can be considered as a reflection of the field of Computer Science itself, considering that it is estimated that only 24% of computer professionals are women. I wrote an article about it:

It's clear that one of the biggest problems here is not the data or the algorithms: it is the blind spots created by the lack of diversity—of experience, education, and thinking—in the teams that develop AI, which makes it challenging to anticipate biases and their potential impact.

Machine learning plays a crucial role in most of the artificial intelligence solutions used today. It is the process by which large amounts of data are used to train AI systems to, for example, extract meaning from text or audio files to answer questions or make recommendations.

Flawed data and biased models can easily lead AI to erroneous conclusions that affect people's credit scores, employment options, school admissions, and their level of risk in criminal court cases.

Artificial intelligence & diversity

Despite being a recent technology, Artificial Intelligence has already shown its value in the field of health, education, and urban mobility, among others.

Considering the pace at AI, Machine Learning, and Deep Learning has been developing in recent years, it is expected that Artificial Intelligence (AI) will profoundly modify the way we live and work.

One of the areas in which this paradigm may stand out in the future is related to Diversity & Inclusion within companies.

AI can detect potential bias and prejudice in decision-making by simulating intelligent behavior, potentially reducing trends and prejudices that could hinder organizations' ability to recruit diversely and inclusively.

Machines do not have an intrinsic propensity to let their personal experiences, opinions, and beliefs influence their decisions. Still, we need to consider that computers are based on data and algorithms created by the people who developed them.

Consequently, if designed and applied ethically, through multidisciplinary and diverse teams, AI can detect situations of potential bias and prejudice in decision making, particularly the one that, being unintentional, becomes more difficult to see.

Can AI itself help us to solve the lack of diversity in AI?

Artificial Intelligence has several ways to impact Diversity & Inclusion within organizations, including promoting equal access to work opportunities and supporting HR departments in making more ethical decisions.

AI can support the development of more diverse and inclusive workplaces allowing access to employment opportunities, supporting the design of more inclusive job ads, managing an ethical recruitment and selection process, and of course, minimizing prejudice during the worker's life cycle.

Despite the AI hype, we must not forget that this technology depends on humans' data collected and selected.

Given that AI solution development depends on human decisions and design, human beings have several unconscious biases; data and machine learning models must be tested and monitored continuously to ensure that they fit their objective in a very ethical way.

By the way, if you want to find out which are your biases, you can take the Implicit Association Test (IAT), carried out by Harvard University.

Conclusion

As important as the improvements and optimizations that AI can bring to society, it must also be put in the condition to be ethical, diverse, and inclusive.

As we saw, the potential benefits of inclusion and diversity are irrefutable; therefore, if automation is a welcome move for companies that invest in technology, the diversity of teams is essential for this to be a path of no return.

Read more about it

If you want to read more about AI ethics, Diversity and Inclusion, and also about volunteering, here are some other articles that I've written about it:

- 14 inspiring and influential women who defy the gender gap in Data Science!
  *Just 15% of today's scientists are women. Like most STEM fields, data science has a daunting gender diversity problem...*medium.com
- The European Commission has its Ethical Guide on Artificial Intelligence. Why does it matter?
  *The European Union hopes that the creation of robust ethical guides will give European technology companies an...*jairribeiro.medium.com
- Here is The Vatican's plan for the development of ethical AI.
  *The Vatican presented a study on how to bring more ethics to the development of artificial intelligence for humanity.*medium.com
- A mini-guide to the E.U.''s new Artificial Intelligence and Data Regulation plan

*European Commission President Ursula von der Leyen wants Europe to have the capability" to create its own choices...*medium.com
- Why do I volunteer, and why should you do it too?
*I started to volunteer with WrOpenUp in 2016, a few months after moving to Wroclaw. I don't remember exactly how it...*medium.com

References:

1. Annual IBM List Celebrates Global Women Leaders Shaping …. https://newsroom.ibm.com/2020-05-06-Annual-IBM-List-Celebrates-Global-Women-Leaders-Shaping-the-Future-of-Artificial-Intelligence
2. Como a falta de diversidade nas equipes de Inteligência Artificial (IA) tem afetado as tecnologias—Patrick Pedreira—https://www.linkedin.com/pulse/como-falta-de-diversidade-nas-equipes-intelig%C3%AAncia-ia-pedreira
3. 'Disastrous' lack of diversity in the AI industry perpetuates …. https://www.theguardian.com/technology/2019/apr/16/artificial-intelligence-lack-diversity-new-york-university-study
4. Corp! salutes diversity award winners—Corp! Magazine. https://www.corpmagazine.com/features/cover-stories/corp-salutes-diversity-winners/
5. Paving the Way for Diversity in the Decade of Ubiquitous AI—https://www.ibm.com/blogs/think/2020/05/paving-the-way-for-diversity-in-the-decade-of-ubiquitous-ai/
6. Racial discrimination persists at Facebook and Google …. https://www.usatoday.com/story/tech/2020/02/10/racial-discrimination-persists-facebook-google-employees-say/4307591002/
7. Charter School Governance: An Exploration of Autonomy and …. https://digitalcommons.georgiasouthern.edu/cgi/viewcontent.cgi?article=2604&context=etd
8. IBM BrandVoice: For AI That Works For Everyone, We Need …. https://www.forbes.com/sites/ibm/2019/12/09/for-ai-that-works-for-everyone-we-need-everyone-to-help-design-it/

9. IIDA's Leadership Discusses the Importance of Diversity in https://www.interiorsandsources.com/article-details/articleid/22798/title/iida-diversity-design-industry
10. Discriminating Systems—Gender, Race, and Power in AI—University of New York—https://ainowinstitute.org/discriminatingsystems.pdf
11. Eye-Opening Statistics on Diversity Every Recruiter Needs .... https://www.censia.com/blog/diversity-statistics/

By Jair Ribeiro on December 7, 2020.

35 - Business vector created by stories - www.freepik.com

What is Prescriptive Analytics, and what can it do for your business.

A great tool that supports businesses, optimizing resources, and increasing operational efficiency.

Prescriptive Analytics is one of the steps of business analytics, including descriptive and predictive analysis. It suggests decision options to take advantage of the results of descriptive and predictive analytics.

It can be utilized to find a solution among various variants, using different simulation and optimization techniques to indicate the path that should be taken.

With Prescriptive Analytics, companies can get smart recommendations to optimize the next steps in their strategy.

Along with predictive analytics, prescriptive analytics help to create a more effective data-based strategy.

Both predictive and prescriptive analytics is critical to making business decisions based on data.

However, the most significant difference between predictive and prescriptive analytics is that predictive analytics predicts what will happen in the future. In contrast, prescriptive analytics offers specific recommendations for changing the future.

With Prescriptive analytics, we can find a solution among various variants to optimize resources and increase operational efficiency. This tool uses different simulation and optimization techniques to indicate the path that should be taken.

Prescriptive analysis is based on:

- Operations investigation
- Predictive Analysis
- Mathematical techniques and statistics

Its application seeks to determine each assumption's limitations based on the study of data and applying mathematical algorithms and probabilistic techniques.

It can be said that it is a learning process that adapts to obtain the best possible result in all real situations that must be faced.

Why prescriptive analysis matters to your business?

Thanks to information obtained through prescriptive analysis, it is possible for companies to make future decisions, such as:

- Calculate past sales of a product to determine the number of replacements.
- Know the tendency of customers in certain products to launch marketing campaigns, according to users' needs.
- Predict equipment failures, which provides for maintenance at the right time.
- Know customers' purchasing habits and punctuality of payment to determine whether it is appropriate to grant credit.

It is possible that some of these decisions can be made manually and correctly. However, the information is bigger and more complicated, and the processes, although more complex, need to be resolved urgently.

Prescriptive analysis has benefits such as:

- Optimization of processes, campaigns, and strategies.
- Minimizes maintenance needs and interconnects them for better conditions.
- Reduce costs without affecting performance.
- It increases the likelihood that companies will approach and plan for internal growth properly.
- Qualitative research method—know the characteristics that distinguish it.
- Production optimization.
- Efficient supply chain management.
- Improved customer service and experience.

Due to its complexity, there are still few companies that use prescriptive analysis.

However, prescriptive analysis benefits have already become evident in many fields, including, but not limited to, healthcare, insurance, financial risk management, and sales and marketing operations.

Among its most significant advantages, it stands out that it allows decision making based on data, Allowing an end-to-end view of costs, processes, and performance.

It is possible to quantify risks and have access to actions considered ideal in different circumstances. Algorithms have the ability, through current data, to predict consequences arising from each decision made.

Therefore, it allows you to follow the path that offers more satisfactory results.

The prescriptive analysis allows more effective planning to be carried out in Marketing and Sales actions, bringing information that significantly impacts business intelligence.

Therefore, decisions are made according to facts, knowing the consequences that will arise from them.

Despite the prescriptive analysis potential, it will only affect a joint work between machine and human being.

That's because technology doesn't make decisions alone. She organizes the information, analyzes the scenario, and indicates the best thing to do, leaving the professional to proceed with the suggestions.

Conclusion

As we saw, Prescriptive Analytics has great potential to support businesses, optimizing resources, and increasing operational efficiency.

They no longer need to use their efforts to analyze data, make projections and research needs, and think of solutions that suggest options that can be used to make future decisions and reduce risk.

This powerful AI tool allows you to process data continuously and improve forecasts to offer new alternatives when making your business decisions.

Other articles you may want to read

If you want to learn more about AI, Machine Learning, and Data Science, I suggest you have a look at these other articles:

- 23 Amazing Youtube Channels for you to Learn AI, Machine Learning, and Data Science for Free...
  This is the perfect moment to start learning something new, and why not start with AI? medium.com
- What is Predictive Analytics, and how can you use it today?
  To see the future, you can rely on two tools: a crystal ball or Predictive Analytics. towardsdatascience.com
- A gentle Introduction to Data Literacy
  Data literacy is a relatively recent trend that simultaneously in business that making informed decisions. Let's find... towardsdatascience.com

By Jair Ribeiro on December 9, 2020.

36 - Business vector created by katemangostar - www.freepik.com

Introduction to the Future with 5G, AI, and IoT.

Artificial Intelligence, IoT, and 5G are set to define our future, but what can we expect from this relationship?

If you were not living in a cave during the last 2 or 3 years, you probably have heard about 5G. AI and IoT.

In many conversations I had, and in many articles I've been reading during the last years, 5G, IoT, and AI has often been associated for many different reasons. And today, I want to explore this enjoyable technological cooperation.

It is more than the internet...

5G is far from being just fast internet. The next generation of mobile telephony is also a paradigm shift in our era. It will allow the use of devices and applications that only work through high data traffic.

In a character of technological revolution, 5G will allow it to become routine to see cars that drive by themselves, doctors that operate at a distance, robots that perform essential tasks in the industry, etc.

According to a recent survey by Qualcomm, already at an economic level and generating revenue of $ 13.2 trillion by 2035, 5G will help create about 22 million jobs by the same year.

The new world scenario with 5G seems like a description of Jetsons' episode. Still, the truth is that this scenario is what awaits us in a short time—and no longer in a distant and hypothetical future.

Getting to know a little bit more about 5G

The first generation (the so-called 1G), it was he who allowed the use of those brick cell phones, a mark of the late 90s and early 2000s.

Then came 2G, the most prominent brand of which was SMS consolidation.

3G arrived soon after. Its differential was internet access, consolidated by 4G, which started to offer a faster internet—and we began to use the cell phone to see series and other actions that required a good connection.

Now it's 5G's turn. Although it is similar to 4G in the sense of needing antennas that transmit electromagnetic waves, to implement a 5G network, it is necessary to build many more structures to handle the waves. Hence, its existence affects the infrastructure of cities.

It is crucial to establish the main differentials of 5G concerning other internet generations:

- Ultra speed: it is estimated that browsing and downloading can be 10 to 20 times faster than we currently know.
- Low latency: do you know that delay when you talk on Skype with someone in another country? It will cease to exist. Today, a delay of 50 milliseconds is calculated; with

5G this number drops to 1 millisecond. This aspect will be critical, for example, to autonomous vehicles soon.

Getting to know a little bit more about AI

Artificial intelligence was created in 1956, but it only became popular today thanks to the increasing volumes of available data, advanced algorithms, and computing power and storage improvements.

The first AI research in the 1950s explored topics such as problem-solving and symbolic methods. It paved the way for the automation and the framework we see in today's computers, including decision support systems and intelligent research systems that can be designed to complement and expand human capabilities.

Over these years, AI has evolved to provide many specific benefits for all industries.

Artificial intelligence is present in machines or devices that reproduce the human mind's functioning in different activities.

These tasks are operated through machine learning, NLP, deep learning, pattern recognition, and sentiment analysis.

More precisely, AI, like Machine Learning, Computer Vision, or Natural Language Processing, is already present in our day, through digital assistants such as Siri, for example; also in e-commerce when using a chatbot to do customer service.

Whenever there is a technology with the capacity to make decisions autonomously, there is an AI behind it.

However, there is much more potential in the relationship between AI and the activities of our day.

And this is where the Internet of Things (or IoT) comes into the game.

Getting to know a little bit more about IoT

The Concept of the Internet of Things, in short, IoT, refers to the connection of various objects to the internet, in addition to those that we are already used to, such as smartphones, tablets, and computers.

These objects, combined with automated systems, can help collect information in real-time, analyze it, and create response actions as needed.

Thus, the Internet of Things is nothing more than an expansion of connectivity.

The search for an understanding of what IoT is and how this innovation will impact our daily lives has intensified.

Imagine that until very recently, a pocket watch marked hours, minutes, and seconds. Today, in addition to the alarm clock, we have smartwatches that make calls, count steps, send messages, carry personal information, and connect to social networks. Seeing the time has become almost a prop.

The idea of IoT is to make things smarter and more connected. Experts say that if you can turn something on and off, it can be connected and part of the IoT universe.

The meaning of IoT, or the internet of things, refers to the set of devices that are not naturally digital but can collect information, send, act on them, or all these actions.

We see IoT in the smart devices we wear daily, like watches that measure user habits. Indoors through devices capable of performing some task, such as voice-activated equipment or a refrigerator with food management tools.

Through a concept called the smart city, cities improve the elements available in the public space. With technology, it is possible to control car traffic, allowing the implantation of vehicle counts in transit and establishing new traffic light configurations.

Finally, IoT is present in the industry, automating some features, improving the supply chain using robots, and reducing errors.

The revolution that is being discussed is represented in this triad: IA, 5G, and IoT: a concept format with three pillars:

- 5G: the fast network that helps applications work.
- Big data: the large volume of data to be processed.
- Artificial intelligence: the algorithms behind smart devices.

What will be the relationship between AI, 5G, and IoT?

The combination of AI, 5G, and IoT will move us to new ways of experiencing the world.

Combined, these technologies will create immense opportunities for users and businesses.

If we consider that in the past, 3G and 4G technologies represented, for people's connectivity, the same as the first steps in the production and use of electricity: the solution of fundamental structural problems we can expect that 5G, together with Artificial Intelligence, will represent a turning point for the entire world economy.

Today, according to a report published by California-based network testing company Viavi Solutions, 5G is at least partially available in 378 cities in a total of 34 countries, and the consumer has been very receptive to technology.

With the adoption of 5G, users will consume, on average, three times more data than the 4G user. It means much more information will be available to the public. This will be a full plate for using technologies that improve consumer understanding like AI, *big data*, and IoT, among others.

As it is a new network that will address more complex problems than previous technologies, the 5G design will benefit from artificial intelligence.

AI will make it possible to predict how customers move around the network and where and at what time they make the most traffic.

Through machine learning, the 5G network will adjust parameters, configuration, and capacity automatically, having a clear benefit from artificial intelligence algorithms and features.

Conclusion

AI, more precisely Machine Learning, and the IoT are part of the world we live in, and very soon, 5G will be so.

It is an important start to pay attention to them to ensure a competitive advantage. Currently, the use of these tools drives businesses to generate automated and more agile services. This consequently impacts the final consumer.

5G will give more power to cloud computing and AI and creates new business possibilities, connecting people in different places around the world as if they were face to face, without delay in communication.

A little bit of confusion between AI and. It exists today because it is challenging to apply Artificial Intelligence to everyday situations without the Internet of Things.

With the enormous amount of data generated by objects connected to the internet, the algorithms make the machines learn and work.

Likewise, the IoT needs Artificial Intelligence to analyze the data collected without a human performing this processing to give immediate and automatic results.

Now the 5G enters this equation with an unforeseen potential for acceleration and disruption.

References

1. Internet of Things (IoT) and Third-Party Risk—Shared .... https://sharedassessments.org/blog/internet-of-things-iot-and-third-party-risk/
2. 5G Economy to Generate $13.2 Trillion in Sales Enablement by 2035— https://www.qualcomm.com/news/releases/2019/11/07/5g-economy-generate-132-trillion-sales-enablement-2035
3. Artificial Intelligence: Big Data and Internet of Things .... https://dig.watch/sessions/artificial-intelligence-big-data-and-internet-things
4. AI and IoT: Tech's New BFFs—InformationWeek. https://www.informationweek.com/big-data/ai-machine-learning/ai-and-iot-techs-new-bffs/a/d-id/1332108
5. Which countries have 5G? | The Week UK. https://www.theweek.co.uk/106957/which-countries-have-5g
6. Artificial Intelligence: Big Data and Internet of Things .... https://dig.watch/sessions/artificial-intelligence-big-data-and-internet-things

By Jair Ribeiro on December 10, 2020.

37 - Business vector created by stories - www.freepik.com

These are some of the best YouTube channels, where you can learn PowerBI and Data Analytics for free.

Power BI is a "powerful" Microsoft program focused on Business Intelligence. You should definitively learn how to use it now.

Power BI is a business analytics service that enables you to see all of your data through a single pane of glass. Live Power BI dashboards and reports include visualizations and KPIs from data residing both on-premises and in the cloud, offering a consolidated view across your business, regardless of where your data lives.

Power BI Desktop is a powerful new visual data exploration and interactive reporting tool. It provides a free-form canvas for drag-and-drop exploration of your data and an extensive library of interactive visualizations while streamlining report creation and publishing to the Power BI service.

It already has its legion of fans (I'm one of them) and an excellent base when it comes to "market share."

Every day more people who already work with BI, Analitycs, Performance, Sales, and even Finance or HR already find in this platform the right solution for the day to day in the company.

One of the main reasons is fluidity. The solution integrates with Microsoft environments (SQL Databases, Azure, DataBricks, Onedrive, Flow, etc....). There is much information about Power BI, the internet being a good part free and relatively complete.

Why PowerBI?

We live in a world where business intelligence goes from an optional tool to a requirement for companies that want to grow. And I believe that in this world, learning Power BI has become essential.

Power BI is one of the most interactive options for businesses that want to start their data-based decision-making path. Its benefits and differentials have given it great popularity among data analysts.

If you want to get valuable insights from your data but you are not yet familiar with the tool or maybe, more than that, if you want to know how to level your team's knowledge in Power BI, here you go.

This article will highlight some of the best YouTube Channels that teach at different levels how to use the tool and even integrate with other solutions with cases close to the real ones found in the professionals' routine.

Microsoft PowerBI Channel (243K subscribers)

First of all, you should have a look at the official PowerBI Youtube Channel by Microsoft.

Microsoft Power BI
Power BI is a business analytics service that enables you to see all

of your data through a single pane of glass.
*Live...*www.youtube.com

PowerBI Tips (6.38K subscribers)

This is the official PowerBI. Tips YouTube channel. Videos and content provided by PowerBI.Tips will be shared here.

Power BI Tips
*This is the official PowerBI.Tips YouTube channel. Videos and content provided by PowerBI.Tips will be shared here.*www.youtube.com

A guy in a Cube (150K subscribers)

A Cube guy is all about helping you master business analytics on the Microsoft Business analytics stack to allow you to drive business growth. The channel looks at how to leverage Microsoft Business Analytics to gain the knowledge needed to shape the data your business cares about. This includes Power BI, Reporting Services, Analysis Services, and Excel. If you work with our business analytics products or services, be sure to subscribe and join in discussion with the weekly content.

Guy in a Cube
*Guy in a Cube is all about helping you master business analytics on the Microsoft Business analytics stack to allow you...*www.youtube.com

BI Elite (28.4K subscribers)

Power BI and DAX tips and tricks from a Microsoft Data Platform MVP. This channel is designed to teach you how to get the most out of Power BI and make you a superuser in no time.

BI Elite
*Power BI and DAX tips and tricks from a Microsoft Data Platform MVP. This channel is designed to teach you how to get...*www.youtube.com

Curbal (57K subscribers)

In this channel, you can learn how to take advantage of your data with Microsoft Power BI and Excel.

Curbal
*Learn how to take advantage of your data with Microsoft Power BI and Excel. We update the channel once a week with: 1...*www.youtube.com

Enterprise DNA (41.6K subscribers)

Enterprise DNA TV has been created for you...the Power BI super users! We are creating an analytical movement to rid the world of poor, time-consuming reporting that makes no value for anyone using this fantastic tool, Power BI. This channel comprehensively covers how to utilize all Power BI areas, with a big focus on using the DAX language to unleash powerful analytical insights from your data.

Enterprise DNA
https://www.youtube.com/c/EnterpriseDNA/featured

RADACAD (13.2K subscribers)

RADACAD is all about helping YOU to get more insight from YOUR data. They publish videos weekly about using Power BI and AI in real-life day-to-day challenges of using these tools to analyze your data and get more meaningful information using dashboards, reports, and visualization in Power BI. And Also, learn how to step up and do predictive analytics using Machine Learning and use AI.

RADACAD
*RADACAD is all about helping YOU to get more insight from YOUR data. We publish videos weekly about using Power BI and...*www.youtube.com

Pavan Lalwani—POWER BI (7.23K subscribers)

Pavan is a freelancer Corporate Trainer for Power BI, Tableau, Microsoft, IBM, and HP software whose mission is to help professionals take control of their skills and present them in a way that inspires, impresses, and builds confidence in their abilities, products, and services.

Pavan Lalwani - POWER BI
*Fascination is one word that describes my curiosity to understand the world around me. "Never let your memories be...*www.youtube.com

Other resources

Not only YouTube channels..., If you want to learn more, the internet is full of free resources—both for those trying to learn from scratch and for those who want to improve their skills on the tool. I would suggest you have a look also at the following:

Microsoft guided learning

What can be better than learning directly from the creators themselves? This free resource from Microsoft allows you to work with the tool to create your first chart to learn advanced features.

With its set of tutorials, you can walk through Power BI very quickly and learn to do some of the most common tasks to get started, including how to ensure data integrity.

It provides an overview of the visualization tool, connecting to data sources, modeling, creating, and customizing simple visualizations. You can also see how to use Excel data in Power BI, an introduction to DAX, and more. You can sign up for free here to take advantage of the resources.

Power BI on Microsoft Learn
*Collections Learn how to get the most out of your organization's dashboards and reports. Explore the collection...*docs.microsoft.com

Microsoft PowerBI blog

This is one of the best ways to learn Power BI and recap all the tool features. It is an excellent addition to your learning. It keeps you updated on new developments that the company brings to the tool directly from the Microsoft Power BI team.

Power BI Blog-Updates and News | Microsoft Power BI
*Today, Microsoft announced the general availability of Azure Synapse Analytics and the preview of Azure Purview, a...*powerbi.microsoft.com

Microsoft PowerBI Webinars

Another essential feature of Microsoft is the webinars, which present several concepts related to Power BI. From introductory training to design concepts, webinars offer videos for the uninitiated and for those who wish to improve their tool skills.

Power BI webinars - Power BI
*Register for our upcoming live webinars or watch our recorded sessions on-demand. Upcoming webinars from the Power BI...*docs.microsoft.com

Official Power BI Community

Microsoft offers a way to dedicate itself to learning Power BI through its interactive community, comprised of technology enthusiasts. In the form of a forum, the community engages discussions on various trending topics about Power BI.

Being a member of this community and participating in lectures can be one of the best ways to equip yourself with the tool's usability and various developments around it. You can encourage your team to participate in the forum.

Microsoft Power BI Community
*This forum is for the students of the EdX.org Power BI class to discuss specific class related questions. If you have a...*community.powerbi.com

Conclusion

Power BI is a powerful tool for managerial vision. However, To get the best from it, you and your team must know how to use the software's full potential. Investing in training in this regard is essential. I hope I have helped you to find the best free resources online.

One more thing...

If you want to read more about Analytics and Data literacy, which are very correlated topics to PowerBI, I've selected some other articles for you:

- What is Predictive Analytics, and how can you use it today?
  *To see the future, you can rely on two tools: a crystal ball or Predictive Analytics.*towardsdatascience.com
- What is Prescriptive Analytics, and what can it do for your business?
  *A great tool that supports businesses, optimizing resources, and increasing operational efficiency.*towardsdatascience.com
- A gentle Introduction to Data Literacy
  *Data literacy is a relatively recent trend that simultaneously in business that making informed decisions. Let's find...*towardsdatascience.com

By Jair Ribeiro on December 15, 2020.

38- Cartoon vector created by vectorjuice - www.freepik.com

What is TinyML, and why does it matter?

Learn the basic concept, the benefits, and where to start into this tiny revolution.

Tiny Machine Learning (or TinyML) is a machine learning technique that integrates reduced and optimized machine learning applications that require "full-stack" (hardware, system, software, and applications) solutions, including machine learning architectures, techniques, tools, and approaches capable of performing on-device analytics at the very edge of the cloud.

TinyML can be implemented in low energy systems, such as sensors or microcontrollers, to perform automated tasks.

With TinyML, we can do more with less. The technique is still ML, but with less energy, costs and without an internet connection.

A small device for a tremendous impact.

This could be a summary for Tiny Machine Learning (or TinyML), emerging breakthroughs within artificial intelligence without exaggeration.

If we consider that, according to a forecast by ABI Research, by 2030, it is likely that around 2.5 billion devices will reach the market through TinyML techniques, having as the primary benefit the creation of smart IoT devices and, more than that, popularize them through a possible reduction in costs.

What's more, the Silent Intelligence consultancy survey reinforces the previous forecast: TinyML can reach more than $ 70 billion in economic value in the next five years. You can't go unnoticed by these figures. Several companies are already organizing themselves to create chips to be used in TinyML implementation.

Also, different ML professionals have been organizing themselves to define this segment's best practices, which is likely to be strengthened very quickly.

Most IoT devices perform a specific task. They receive input via a sensor, perform calculations, and send data or perform an action.

The usual IoT approach is to collect data and send it to a centralized registration server. Then, you can use machine learning to conclude.

But why don't we make these devices smart at the embedded system level? We can build solutions like smart traffic signs based on traffic density, send an alert when your refrigerator runs out of stock, or even predict rain based on weather data.

The challenge with embedded systems is that they are tiny. And most of them run on battery. ML models consume a lot of processing power. Machine learning tools like Tensorflow are not suitable for creating models on IoT devices.

Cracking the small ML

In TinyML, the same ML architecture and approach is used, but on smaller devices capable of performing different functions, from answering audio commands to executing actions through chemical interactions.

But how do we get TinyML? Many tools can help us to run machine learning models on IoT devices.

The most famous is Tensorflow Lite. With Tensorflow Lite, you can group your Tensorflow models to run on embedded systems. Tensorflow Lite offers small binaries capable of running on low power embedded systems.

One example is the use of TinyML in environmental sensors. Imagine that the device is trained to identify temperature and gas quality in a forest. This device can be essential for risk assessment and identification of fire principles.

Some of the main differentials of the technology are:

- Data Security: As there is no need to transfer information to external environments, data privacy is more guaranteed.
- Energy savings: Transferring information requires an extensive server infrastructure. When there is no data transmission, energy and resources are saved, consequently in costs.
- No connection dependency: If the device depends on the Internet to work, and it goes down, it will be impossible to send the data to the server. You try to use a voice assistant, and it does not respond because it is disconnected from the Internet.
- Latency: Data Transfer takes time and often brings in a *delay*. When it does not involve this process, the result is instantaneous.

Python is generally the preferred language for building ML models. Still, with TensorFlow Lite, you can use C, C ++, or Java to create machine learning models.

Connecting to the network is an energy-consuming operation. Using Tensorflow Lite, you can deploy machine learning models without the need to connect to the Internet. This also solves security issues since embedded systems are relatively easier to exploit.

Tensorflow Lite offers pre-trained machine learning models for everyday use cases. These include:

- Object detection is used to recognize multiple objects in an image, supporting up to 80 different items.
- Smart responses—Generates intelligent responses, similar to what you get when interacting with a conversational AI or a chatbot.
- Recommendations—It offers customized recommendation systems based on user behavior.

There are some valid alternatives to Tensorflow Lite. Two strong competitors are:

- CoreML—Apple library for building machine learning models on iOS devices.
- PyTorch Mobile—mobile version of Facebook's PyTorch deep learning library.

TinyML is still in its early stages. Improvements are being made to Tensorflow Lite and other TinyML frameworks to support complex machine learning models.

It may take some time before we definitively start to see the dominant adoption of TinyML. But make no mistake, smart devices are coming.

Where can you learn more about TinyML?

Currently, the leading community is around the tinyML Foundation, which aims to build a global community of researchers, engineers, product managers to develop cutting-edge technology, promoting and stimulating knowledge on the subject.

But I would like to recommend a fascinating book (I'm reading it at this moment, and probably I will write a review about it very soon..) called *Tiny ML: Machine Learning with Tensorflow Lite on Arduino and Ultra-Low-Power Microcontrollers,* by Pete Warden and Daniel Situnayake, that is an introductory work to the TinyML universe.

The book aims to help understand how we can train small models who understand audio, image, and data to perform some tasks. According to the book's description, no previous ML or microcontrollers' experience is necessary to accompany the work. I think it worths to have a look.

Conclusion

TinyML will open up a series of possibilities for applications in IoT devices, such as TVs, cars, coffee machines, watches, and other devices, so that they have intelligent functionalities that today are restricted in computers and smartphones.

We will see voice interfaces in almost everything in the future. As soon as we can create suitable voice interfaces at a low cost, we will have them on any consumer item, replacing buttons on any devices, especially if you think of devices combining audio and video. I want to be ready for that, what about you?

One more thing...

If you want to read more about Machine Learning, AI, IoT, and 5G, I've selected some interesting articles that you may like to read:

- Introduction to the Future With 5G, AI, and IoT.
  *Artificial Intelligence, IoT, and 5G are set to define our future, but what can we expect from this relationship?*medium.com
- The best free courses to learn AI, ML, and Data Science today.
  *More than 60 courses with ratings and a brief summary (Made by AI, of course).*medium.com
- The most impressive Youtube Channels for you to Learn AI, Machine Learning, and Data Science.
  *This is the perfect moment to start learning something new, and why not start with AI?*medium.com

References

1. Google Scholar—TinyML—https://scholar.google.com/scholar?hl=en&as_sdt=0%2C5&q=tinyML&btnG=
2. MicroNets: Neural Network Architectures for Deploying TinyML Applications on Commodity Microcontrollers—https://arxiv.org/abs/2010.11267
3. TensorFlow Lite Micro: Embedded Machine Learning on TinyML Systems—https://arxiv.org/abs/2010.08678
4. Why the Future of Machine Learning is Tiny—https://petewarden.com/2018/06/11/why-the-future-of-machine-learning-is-tiny/
5. How Engineers Are Using TinyML to Build Smarter Edge Devices—https://new.engineering.com/story/how-engineers-are-using-tinyml-to-build-smarter-edge-device
6. tinyml如何使用TensorFlow Lite构建智能物联网设备_weixin_26750481的博客-CSDN博客. https://blog.csdn.net/weixin_26750481/article/details/108499905
7. Why TinyML is a giant opportunity—https://venturebeat.com/2020/01/11/why-tinyml-is-a-giant-opportunity/

By Jair Ribeiro on December 22, 2020.

39 - Business vector created by vectorjuice - www.freepik.com

The Best MIT Online Resources for You to Learn AI and Machine Learning for Free

Studying at MIT can be very expensive, but currently, more than 200 courses are available for free, and here you have a list of some of the...

You probably heard about the Massachusetts Institute of Technology, also known as MIT.

MIT is one of the world's leading centers of study and research in science, engineering, and technology. Founded in 1861 in Cambridge, USA, the institute trained professionals to meet the demand of industries growing fast. Only in the mid-1930s did MIT focus its training on basic scientific research and technological innovation.

Studying at MIT is expensive, around $ 60,000 a year, and its average scholarship amount for international students is the US $ 32,000—but in some cases, it can reach 100%. On average, 62% of its students receive some form of financial aid. In graduate school, almost 90% of students receive some scholarship. But there is another way to be part of it?

More than 200 courses are available—many of them in science and technology, but with options also in economics, business, history, biology, sociology, and others.

Since 2008, the institution started to incorporate videos into online courses. Currently, more than 100 courses have complete video classes with teachers from the institution. But our focus here will be AI and Machine Learning courses.

Why should you learn AI and Machine Learning?

Artificial Intelligence (AI) and machine learning—its main component today—are two of the most recurring themes today regarding innovation. Although most of the approaches are quite rich and positive, there is still a little exaggeration in the results' expectations. And sometimes even suggestions for applications where AI would not necessarily be the best option.

Artificial intelligence is not a trend. It is a fact that there is still a little exaggeration in forecasts for the future.

However, a recent cycle of advances in algorithms and computational infrastructure has generated commercially relevant results.

Today, even professionals with other backgrounds, such as *designers*, developers, *testers, and product own*ers, can and should have a broad view of AI.

In this way, they can think about the day-to-day, in each of their projects, how to incorporate these concepts, advising customers about it, and generating more value for all involved.

I believe that many professionals outside the exact sciences can and should learn the basics of AI to increase their potential contribution, either for their employers or our society.

Are you ready?

But before you jump into the MIT online learning experience, I would like to invite you to search and watch on YouTube the exciting lecture by professor Jeremy Kepner and Vijay Gadepally from the MIT Lincoln Laboratory that will provide an overview of artificial intelligence and took a deep dive into machine learning, including supervised learning, unsupervised learning, and reinforcement learning.

Now you can go to and find the best training course by MIT that will help you to build your deep expertise in AI and Machine Learning:

Artificial Intelligence

This course introduces students to the basic knowledge representation, problem-solving, and learning methods of artificial intelligence. Upon completion of 6.034, students should be able to develop intelligent systems by assembling solutions to concrete computational problems; understand the role of knowledge representation, problem-solving, and learning in intelligent-system engineering; and appreciate the role of problem-solving, vision, and language in understanding human intelligence from a computational perspective. [1]

Artificial Intelligence
*This course introduces students to the basic knowledge representation, problem solving, and learning methods of...*ocw.mit.edu

Introduction to Machine Learning

This course explores simulation and prediction concepts, algorithms, and machine learning applications. It involves formulating learning problems, representation definitions, overfitting, and generalization. These principles are exercised in supervised learning and reinforcement learning, applying to images and temporal sequences.

Introduction to Machine Learning
This course introduces principles, algorithms, and applications of

machine learning from the point of view of modeling...openlearninglibrary.mit.edu

Machine Learning

This course introduces machine learning that provides an overview of many machine learning principles, techniques, and algorithms, starting with classification and linear regression and concluding more recent issues like boosting, supporting vector machines, and hidden Markov models Bayesian networks. The course will give students the fundamental ideas and intuition behind current machine learning methods and a more systematic understanding of how, why, and when they work. The underlying theme is statistical inference since it provides the basis for most of the methods covered.

Machine Learning
6.867 is an introductory course on machine learning which gives an overview of many concepts, techniques, and...ocw.mit.edu

Introduction to Computational Thinking and Data Science

This course is for learners with little to no programming experience. It aims to offer students an understanding of computing's role in solving problems and helping students write small programs.

Using Python 3.5 programming language.

Introduction to Computational Thinking and Data Science
6.0002 is the continuation of 6.0001 Introduction to Computer Science and Programming in Python and is intended for...ocw.mit.edu

Techniques in Artificial Intelligence (SMA 5504)

A graduate-level guide to artificial intelligence. Topics discussed include first-order representation and inference, current deterministic and decision-theoretical planning techniques, simple supervised learning methods, and Bayesian network inference and learning.

Techniques in Artificial Intelligence (SMA 5504)
*6.825 is a graduate-level introduction to artificial intelligence. Topics covered include: representation and inference...*ocw.mit.edu

Mathematics of Machine Learning

Machine Learning refers to automatic pattern recognition in data. As such, new statistical and algorithmic advances became the fertile ground. This course aims to provide a mathematically rigorous introduction to these developments with emphasis on methodology and interpretation.

Mathematics of Machine Learning
*Broadly speaking, Machine Learning refers to the automated identification of patterns in data. As such it has been a...*ocw.mit.edu

Introduction to Deep Learning

MIT's introductory course on deep learning approaches with computer vision, natural language processing, genetics, and more! Students gain basic knowledge of deep learning algorithms and experience in developing neural networks in TensorFlow. The course ends with a project plan competition, input from staff, and industry sponsors panel. Preconditions presume calculus (i.e., derivatives) and linear algebra (i.e., matrix multiplication)

Introduction to Deep Learning
*This is MIT's introductory course on deep learning methods with applications to computer vision, natural language...*ocw.mit.edu

Machine Learning for Healthcare

This course introduces students to healthcare machine learning, including the essence of clinical data and machine learning techniques for risk stratification, disease progression modeling, precise medicine, diagnosis, subtype discovery, and clinical workflow enhancement.

Machine Learning for Healthcare
Peter Szolovits, and David Sontag. 6.S897 Machine Learning for Healthcare. Spring 2019. Massachusetts Institute of... ocw.mit.edu

Deep Learning for Self Driving Cars

As with the course above, MIT uses a significant real-world feature of AI as a jump-off point to explore the particular technologies involved.

The self-driving cars widely expected to become part of our daily lives rely on AI to make sense of all data hitting the vehicle's sensor array and safely navigating the roads. This includes training computers to interpret sensor data just as our brains interpret eye, ear, and touch signals.

It will teach you how to use the MIT DeepTraffic simulator, which challenges students to teach a virtual car to travel along a busy road as quickly as possible without colliding with other pedestrians.

This is a course taught at the bricks' n' mortar university for the first time last year, and all resources, including lecture videos and exercises, are available online—but you won't get a certificate.

MIT Deep Learning and Artificial Intelligence Lectures | Lex Fridman
A collection of lectures on deep learning, deep reinforcement learning, autonomous vehicles, and artificial... deeplearning.mit.edu

Enter Julia, a rising start on AI development.

Frederik Bussler recently wrote an interesting article about MIT's training course that will teach you Julia, the rising star among the programming languages, that could replace python as the favorite language for AI and Machine Learning development.

In the article, Frederik mentions that MIT recently announced a free online course on computational thinking, taught using Julia. I think you should look into Introduction to Computational Thinking with Julia, with Applications to Modeling the COVID-19 Pandemic. A half-semester course introduces computational thinking through data science applications, artificial intelligence, and mathematical models using the Julia programming language. This Spring 2020 version is a fast-tracked curriculum adaptation to focus on applications to COVID-19 responses.

Introduction to Computational Thinking with Julia, with Applications to Modeling the COVID-19...
This half-semester course introduces computational thinking through applications of data science, artificial... ocw.mit.edu

Discovering a brave new planet

The second bonus is not a training course, but it is a podcast by MIT that caught my attention this week:

Hosted by scientist Dr. Eric Lander, president and founding director of the Broad Institute of MIT and Harvard. He is a geneticist, molecular biologist, and mathematician who led the Human Genome Project and served as President Obama's White House science advisor for eight years.

Brave New Planet is a podcast about remarkable new technologies that might dramatically improve our world, or if we don't make wise choices, it could leave us a lot worse off. It delves deep into the most exciting and challenging scientific frontiers, helping us understand them and grapple with their implications.

Homepage | Brave New Planet
Hosted by scientist Eric Lander and in partnership with Pushkin Industries and the Boston Globe, Brave New Planet is a... www.bravenewplanet.org

Conclusion

Through these fantastic pieces of training, you will have the opportunity to develop your skills about how artificial intelligence and machine learning methods work under various circumstances. Of course, you will understand and appreciate the relevance of problem-solving, vision, and language in understanding human intelligence from the AI and Machine Learning perspective.

Being part of MIT (the online experience is very inclusive), you'll be part of a community of restless learners, enthusiastic dreamers, and extraordinary doers. And with this list of training I've shared here, you get all this for free

Read more about it...

If you still are here, probably you would like to read more about where you can learn more AI, ML, and Data Science courses and about my experience as a MicroMasters's student at MIT:

- I've just started a (micro)Masters at the MIT
  *"The funniest part is to start a Masters; the hardest part is to finish it"—myself.*medium.com
- The most impressive Youtube Channels for you to Learn AI, Machine Learning, and Data Science.
  *This is the perfect moment to start learning something new, and why not start with AI?*medium.com
- These are some of the best Youtube channels where you can learn PowerBI and Data Analytics for...
  *Power BI is a "powerful" Microsoft program focused on Business Intelligence and you should definitively learn how to...*towardsdatascience.com
- The best free courses to learn AI, ML, and Data Science today.
  *More than 60 courses with ratings and a brief summary (Made by AI, of course).*medium.com

By Jair Ribeiro on December 23, 2020.

References

- Maybe one day, a robot will steal your job ... but there is https://medium.com/predict/maybe-one-day-a-robot-will-steal-your-job-but-there-is-something-you-can-do-today-to-avoid-that-ad75486e6199
- Will My Daughters Ever Drive a Car? | by Jair Ribeiro https://medium.com/swlh/will-my-daughters-never-drive-a-car-6e579158717a
- A mini-guide to the E.U.''s new Artificial Intelligence https://medium.com/tech-cult-heartbeat/a-mini-guide-to-the-e-u-s-new-artificial-intelligence-and-information-regulation-plan-487e263f8e8d
- Technology, Culture, and Real-life stories – Medium. https://medium.com/tech-cult-heartbeat
- Top 25 Data Science Influencers to Follow in 2020. https://roundtable.datascience.salon/top-25-data-science-influencers-to-follow-in-2020
- We should learn how to collaborate with robots before it https://medium.com/tech-cult-heartbeat/you-must-learn-how-to-collaborate-with-robots-before-its-too-late-fed039aec0eb
- Introduction to the Future With 5G, AI, and IoT. | by Jair https://medium.com/swlh/introduction-to-the-future-with-5g-and-artificial-intelligence-10239bd452a1
- 3 points to consider if you think that AI will take your https://medium.com/predict/3-points-to-consider-if-you-think-that-ai-will-take-your-job-c54b24bb9c91
- Ethics and Artificial Intelligence | Connessioni .... https://www.connessioni.biz/en/ethics-and-artificial-intelligence/
- Summarizing A.I. Articles using A.I. (What else??) | by https://medium.com/towards-artificial-intelligence/summarizing-ai-articles-using-ai-what-else-c1c16a80dbb5
- Thinkers360. https://www.thinkers360.com/tl/jairribeiro
- Pope Francis Offers 'Rome Call For AI Ethics' To Step-Up https://www.forbes.com/sites/lanceeliot/2020/03/10/pope-

- francis-offers-rome-call-for-ai-ethics-to-step-up-ai-wokefulness-which-is-a-wake-up-call-for-ai-self-driving-cars-too
- Three Intelligent Apps that Make you more Productive Today https://medium.com/towards-artificial-intelligence/3-intelligent-apps-that-make-you-more-productive-today-2e89d2e32707
- The Catholic Church and Pope Francis plan to fight back https://www.vox.com/recode/2020/2/28/21157760/pope-vatican-artificial-intelligence
- Colorizing Black & White images with Deep Learning | by https://medium.com/towards-artificial-intelligence/colorizing-images-with-deep-learning-a34d11587643
- An easy guide to the history of Artificial Intelligence https://medium.com/tech-cult-heartbeat/an-easy-guide-to-the-history-of-artificial-intelligence-37a07a1ad238
- Artificial Intelligence: EU wants to develop Ethical Rules https://www.elektroniknet.de/international/eu-wants-to-develop-ethical-rules-for-ai-164251.html
- How BlueDot anticipated Coronavirus using Artificial .... https://www.techaheadcorp.com/blog/how-bluedot-anticipated-coronavirus-using-ai
- Vatican signs up IBM and Microsoft as AI ethics apostles https://devclass.com/2020/03/02/vatican-signs-up-ibm-and-microsoft-as-ai-ethics-apostles/
- God - Center for Mission. https://centerformission.org/wp-content/uploads/2020/11/November-20.pdf
- Thoughts About the Need for More Diversity and Inclusion https://medium.com/swlh/the-need-for-more-diversity-and-inclusion-in-artificial-intelligence-f78e398282c
- A Simple Approach to Define Human and Artificial https://medium.com/towards-artificial-intelligence/a-simple-approach-to-define-human-and-artificial-intelligence-4d91087d16ff
- An easy guide to the history of Artificial Intelligence https://medium.com/tech-cult-heartbeat/an-easy-guide-to-the-history-of-artificial-intelligence-37a07a1ad238

- The Best MIT Online Resources for You to Learn AI and https://medium.com/swlh/the-best-mit-online-resources-for-you-to-learn-ai-and-machine-learning-for-free-d3ba1e50f436
- MIT Artificial Intelligence | 23 Lectures | Patrick H https://www.newworldai.com/artificial-intelligence-complete-lectures-01-23/
- The most impressive Youtube Channels for you to Learn AI https://medium.com/swlh/21-amazing-youtube-channels-for-you-to-learn-ai-machine-learning-and-data-science-for-free-486c1b41b92a
- These are some of the best Youtube channels where you can https://towardsdatascience.com/the-top-youtube-channels-for-you-to-learn-powerbi-and-data-analytics-for-free-8f8eb434b48d
- 5 companies that are revolutionizing recruiting using https://towardsdatascience.com/5-companies-that-are-revolutionizing-recruiting-using-artificial-intelligence-9a70986c7a7e
- Essential Hard & Soft Skills to Pursue in 2020: LinkedIn https://www.hrinasia.com/people-development/%EF%BB%BFessential-hard-soft-skills-to-pursue-in-2020/
- Stop everything you are doing and watch these 5 TED Talks https://medium.com/tech-cult-heartbeat/stop-everything-you-are-doing-and-watch-these-5-ted-talks-on-ai-ethics-now-aaa865f499fb
- Blockchain was the most in-demand job skill in 2020, says https://www.cnbc.com/2020/01/17/blockchain-is-the-most-in-demand-job-skill-in-2020-says-linkedin.html
- Reinforcement Learning and 9 examples of what you can do https://towardsdatascience.com/about-reinforcement-learning-2ff0dafe9b75
- 12 (+Bonus) amazing Youtube Channels To Learn Python https://medium.com/towards-artificial-intelligence/21-great-youtube-channels-for-you-to-learn-python-programming-for-free-d6470c444f7d
- Maybe one day, a robot will steal your job ... but there is .... https://medium.com/predict/maybe-one-day-a-robot-

- will-steal-your-job-but-there-is-something-you-can-do-today-to-avoid-that-ad75486e6199
- LinkedIn reveals the most in-demand job skills in 2020 .... https://www.cnbcafrica.com/financial/2020/01/27/linkedin-reveals-the-most-in-demand-job-skills-in-2020
- How I am summarizing the most relevant terms of Artificial https://medium.com/dataseries/how-i-am-summarizing-the-most-relevant-terms-of-artificial-intelligence-from-wikipedia-using-ai-7e85314d81bf
- Europe leads the way on set rules for Artificial https://medium.com/the-innovation/europe-leads-the-way-on-set-rules-for-artificial-intelligence-7b4d734a8595
- How AI and Digital Transformation will change your https://towardsdatascience.com/how-ai-and-digital-transformation-will-change-your-business-forever-c7563c15c1b3
- How Autonomous Vehicles will redefine the concept of https://towardsdatascience.com/how-autonomous-vehicles-will-redefine-the-concept-of-mobility-582f8701a5f8
- Surveillance capitalism: Who is watching us online — and https://www.cbc.ca/radio/ideas/surveillance-capitalism-who-is-watching-us-online-and-why-1.5791546
- What is TinyML, and why does it matter? | by Jair Ribeiro https://jairribeiro.medium.com/what-is-tinyml-and-why-does-it-matter-f5b164766876
- What is Lobe and how is Microsoft Trying to Make AI https://medium.com/towards-artificial-intelligence/enter-lobe-and-how-microsoft-is-trying-to-make-ai-mainstream-419c9dfe55f5
- Enter Lobe and how Microsoft is trying to make AI https://laptrinhx.com/enter-lobe-and-how-microsoft-is-trying-to-make-ai-mainstream-2313398211/
- Joy Buolamwini - Doha Debates. https://dohadebates.com/participant/joy-buolamwini/
- Overview ‹ Joy Buolamwini — MIT Media Lab. https://www.media.mit.edu/people/joyab/overview
- The Skills Companies Needed Most in 2020—And How to Learn https://campusalliedhealth.com/blog/the-skills-companies-need-most-in-2020-and-how-to-learn-them

- An Introduction to Autonomous Vehicles | by Jair Ribeiro https://towardsdatascience.com/an-introduction-to-autonomous-vehicles-b39024788cd6
- Rebooting Regulation: May 2019 Exploring the Future of AI .... https://brookfieldinstitute.ca/wp-content/uploads/AIFutures_PolicyLabs_Final_EN.pdf
- Joy Buolamwini - Doha Debates. https://dohadebates.com/participant/joy-buolamwini/
- Thoughts About the Need for More Diversity and Inclusion https://medium.com/swlh/the-need-for-more-diversity-and-inclusion-in-artificial-intelligence-f78e398282c
- Blockchain was the most in-demand job skill in 2020, says https://www.cnbc.com/2020/01/17/blockchain-is-the-most-in-demand-job-skill-in-2020-says-linkedin.html
- photograph | AITopics. https://aitopics.org/tag/photograph
- Renata Rawlings-Goss | Covid Information Commons. https://covidinfocommons.datascience.columbia.edu/content/renata-rawlings-goss
- Fiddler Uses AWS to Makes it Easy for Companies to Explain .... https://aws.amazon.com/blogs/startups/fiddler-uses-aws-to-makes-it-easy-for-companies-to-explain-ml-models
- The Skills Companies Needed Most in 2020—And How to Learn https://campusalliedhealth.com/blog/the-skills-companies-need-most-in-2020-and-how-to-learn-them
- Vatican Inks Artificial Agreement with High-Tech Companies .... https://saintcharles.co.za/vatican-inks-artificial-agreement-with-high-tech-companies-zenit/
- Rome Call for AI Ethics - Innovation Post. https://innovationpost.it/wp-content/uploads/2020/02/Rome-Call-for-AI-Ethics.pdf
- Can "fake faces" lead to the illusion of diversity? | by https://medium.com/tech-cult-heartbeat/fake-faces-fake-diversity-1798f96c7371
- Current: The Skills Companies Need Most in 2020 .... https://executivewomen20.com/2020/01/14/current-the-skills-companies-need-most-in-2020
- The Ethics of AI and Autonomous Vehicles | by Jair Ribeiro .... https://medium.com/towards-artificial-

intelligence/the-ethics-of-ai-and-autonomous-vehicles-5ddfa0fa2726
- Honda will be the first to bring autonomous vehicles to https://medium.com/the-innovation/honda-will-bring-the-level-3-autonomous-vehicles-to-the-masses-747ae9385105
- The Top Skills In Demand For 2020—And How to Learn Them. https://www.linkedin.com/business/learning/blog/top-skills-and-courses/the-skills-companies-need-most-in-2020and-how-to-learn-them
- Meet the 2020 Women in Tech Initiative Athena Awards Winners. https://citris-uc.org/meet-the-2020-witiuc-athena-awards-winners
- Current: The Skills Companies Need Most in 2020 .... https://executivewomen20.com/2020/01/14/current-the-skills-companies-need-most-in-2020/
- What is Predictive Analytics, and how can you use it today https://towardsdatascience.com/what-is-predictive-analytics-dc6db9759936
- 5 amazing books about AI that you must read in 2020. | by https://towardsdatascience.com/amazing-books-about-ai-that-you-should-be-reading-4db9646e0807
- iHarare Jobs - The Skills Companies Needed Most in 2020—And https://ihararejobs.com/job/the-skills-companies-need-most-in-2020-and-how-to-learn-them
- The Top Skills In Demand For 2020—And How to Learn Them. https://www.linkedin.com/business/learning/blog/top-skills-and-courses/the-skills-companies-need-most-in-2020and-how-to-learn-them

Table of Contents

Dedication 2

Epigraph 3

Acknowledgments 4

Preface 5

Maybe one day, a robot will steal your job… but there is something you can do today to avoid that. 7

    The leaders of the future 9

    So, what do you can do to keep your work in the future? 9

Stop everything you are doing and watch these five TED Talks on A.I. Ethics now. 11

    How to Keep Human Bias Out of AI 12

    Can We Protect A.I. from Our Biases? 13

    The Era of Blind Faith in Big Data Must End 13

    Machine Intelligence Makes Human Morals More Important 14

    How to Get Empowered Not Overpowered 15

    It's time for action 15

An easy guide to the history of Artificial Intelligence 17

Thirty-eight free courses to help you master the most in-demand job skills in 2020. 24

    Not only Hard skills… 30

Can "fake faces" Lead to the Illusion of Diversity? 34

    A business opportunity 34

    Inverting the roles: Not only A.I. faces but also A.I. beauty contests 36

    Is it possible to use GANs for good today? 37

    Fighting bias with GANs 38

    Solving privacy issues with Synthetic data. 38

    Conclusion 38

Can Artificial Intelligence predict the next pandemic? 40

    But could this epidemic outrage be predicted? 40

    A.I. as a disruptive wave in healthcare 41

    Conclusion 42

Three intelligent apps that make you more productive today 43

    Calendar 44

    Habit—Daily Tracker 45

- Things 45
- Bonus app: Grammarly 47
- Conclusion 47

The European Commission has its Ethical Guide on Artificial Intelligence. Why does it matter? 48
- Building a trustworthy A.I. 49
- The seven principles 49
- Why do we need an open debate about A.I.? 50

We should learn how to collaborate with robots before it is too late. 51
- Automation and the future of work 52
- Entering the era of integration between Humans and Machines 52
- Technical capacity: we must understand how machines work and continuous learning about them. 52
- Data capacity: we must learn how to analyze and interpret information generated by machines. 53
- Human capacity: developing the fundamental human capabilities that machines cannot imitate, or soft skills. 53
- How to cooperate and collaborate with machines at work 54

Three facts to consider if you think that A.I. will take your job. 56
- A.I. and the future of jobs 56
- Job Created by Artificial Intelligence 57
- What's the future between Humanity and the Machines? 58

A mini-guide to the E.U.' 's new Artificial Intelligence and Data Regulation plan 59
- Artificial Intelligence is the future. 61
- No facial recognition ban, for now. 61
- Data is the new oil for Europe. 62
- Europe also wants to assess the potential risks of Artificial Intelligence. 63
- Conclusion 63

Five amazing books about AI that you should be reading 64
- The Book of Why: The New Science of Cause and Effect 65
- How to Create a Mind: The Secret of Human Thought Revealed 66
- Life 3.0: Being Human in the Age of Artificial Intelligence 66
- The Emotion Machine: Commonsense Thinking, Artificial Intelligence, and the Future of the Human Mind 67
- Artificial Intelligence: A Modern Approach 68

- Conclusion ... 69
- Here is The Vatican's plan for the development of ethical AI. ... 70
  - Towards an "algorithmic ethics." ... 71
  - The prospect of a good AI ... 72
  - For an ethical AI development ... 72
  - Conclusion ... 73
- How I am summarizing the most relevant terms of Artificial Intelligence from Wikipedia, using AI. ... 74
  - Why Text summarization matter? ... 75
  - Different approaches to Text Summarization ... 76
  - Steps involved to create the text summary ... 76
  - Summarizing the articles. ... 78
    - How I collected the data? ... 78
- Summarizing A.I. Articles using A.I. ... 79
  - Article Summarization tool using NLP ... 79
  - Why Text summarization matter? ... 80
  - Different approaches to Text Summarization ... 80
  - Steps involved to create the text summary ... 81
  - Using Transfer Learning for summarizing the articles ... 82
- Colorizing Black & White images with Deep Learning ... 84
  - The Deep Learning approach. ... 85
  - Working with the LAB ... 86
  - Conclusion ... 87
  - Links and Sources: ... 88
- The Ethics of AI and Autonomous Vehicles ... 89
  - From Asimov and beyond ... 90
  - Some examples of ethical issues and self-driving vehicle dilemmas ... 91
  - What about children and autonomous vehicles? ... 92
  - Are humans predictable? ... 93
  - The AI ethics issues and our responsible future. ... 94
  - Conclusion ... 94
- An Introduction to Autonomous Vehicles ... 96
  - We will be driving in a better world. ... 97
  - Learning by doing: Home-made autonomous driving systems ... 97

- What are autonomous vehicles? 98
- Not every Autonomous vehicle is made equal: the six levels of autonomy. 99
- How do autonomous vehicles work in practice? 100
- The technology behind autonomous cars? 101
- The Hardware 101
  - Stereoscopic camera 101
  - Infrared camera 102
  - Radar 102
  - Sonar 102
  - LIDAR 102
  - ESC (Electronic Stability Control) 102
  - iBooster 102
  - GPS, speedometer, and odometer 103
- The software 103
  - Perception 103
  - Planning 103
  - Control 104
- Artificial Intelligence and Connectivity 104
- Vehicles that can see the world 105
- Do you foresee the unusual 105
- The Autonomous driving revolution 105
- Conclusion 106
- Autonomous Vehicles will redefine the concept of mobility 108
  - A disruptive wave: Transportation as a Service 109
  - With more significant autonomy comes more significant challenges 111
  - Autonomous vehicles and the new customer experience 111
  - The impact of Autonomous vehicles on automakers 112
  - About maintenance of Autonomous vehicles 113
  - What changes with the arrival of autonomous vehicles on the transport industry chain? 114
    - Spare parts market 114
    - Service stations 115
    - Mechanical workshops 115
    - Vehicle trades 115

    Conclusion 116
Fourteen inspiring and influential women who defy the gender gap in Data Science! 117
    A career disconnection 118
    We need more women in Data Science. 119
    An inspiring list of women in Data Science 120
        Fei-Fei Li 120
        Cassie Kozyrkov 121
        Allie Miller 121
        Elizabeth M. Adams 122
        Tamara McCleary 123
        Carla Gentry 123
        Danielle Belgrave 124
        Kristen Kehrer 124
        Joy Buolamwini 125
        Chip Huyen 125
        Fay Cobb Payton 126
        Kate Crawford 126
        Dr. Renata Afi Rawlings-Goss 127
    Conclusion 128
    Read more about it… 128
What is Reinforcement Learning, and nine examples of what you can do with it? 130
    Reinforcement Learning Challenges 131
    Applications areas of Reinforcement Learning 132
        Games 132
        Personalized Recommendations 132
        Resource Management in Computer Clusters 133
        Traffic Light Control 133
        Robotics 134
        Web Systems Configuration 134
        Chemistry 134
        Auctions and Advertising 135
        Deep Learning 135
    Conclusion: When should you use RL? 136

- Resources 137
- What is Lobe, and how is Microsoft Trying to Make AI mainstream? 138
  - What is Lobe? 139
  - What can you do with Lobe? 139
  - Conclusion 140
  - Read more about it… 140
- The most impressive YouTube Channels for you to Learn AI, Machine Learning, and Data Science. 141
  - SpringBoard 142
  - Arxiv Insights 142
  - Machine Learning 101 143
  - FreeCodeCamp 143
  - Data School 143
  - Machine Learning TV 144
  - Giant Neural Network 144
  - Andreas Kretz 144
  - Edureka! 145
  - Andrew Ng 145
  - Deeplearning.ai 145
  - Tech with Tim 146
  - Machine Learning University (MLU) 146
  - Artificial Intelligence—All in One 147
  - Sentdex 147
  - Joma Tech 147
  - Python Programmer 148
  - Deep Learning TV 148
  - Google Cloud Platform 148
  - Keith Galli 149
  - Data Science Dojo 149
  - Updates from our readers 149
    - TechnoBotic 149
    - StatQuest 150
    - Yannic Kilcher 150
  - Conclusion 150

- Read more about it… 150
- Europe leads the way on set rules for Artificial Intelligence 152
  - An ethics framework for AI 153
  - Liability for AI causing damage 153
  - Intellectual property rights 154
  - Conclusion 154
  - More information: .... **Error! Bookmark not defined.**
- A gentle Introduction to Data Literacy 155
  - What is Data Literacy 155
  - The Data Literacy Project 157
  - The impact of Data Literacy 158
  - Conclusion 159
- How AI and Digital Transformation will change your business forever. 161
  - Defining the Digital Transformation scenario 162
  - Before talking about AI, do you speak data? 163
  - Artificial Intelligence and the Digital Transformation 163
  - Where to start? 164
  - Defining your journey to Artificial Intelligence 165
  - How can you create value with Digitization through Artificial Intelligence? 167
  - The importance of an early adoption 167
  - Conclusion 168
- Honda will bring the level three autonomous vehicles to the masses. 170
  - Autonomy Levels 171
  - Read more about it… 172
- Five companies that are revolutionizing recruiting using Artificial Intelligence 174
  - Some applications of artificial intelligence in recruitment and selection ... 175
  - Five disruptive companies that are revolutionizing recruiting using AI 176
    - HireVue 176
    - Mya Systems 176
    - HiredScore 176
    - Wade & Wendy 177
    - Hiretual 177
  - Conclusion 178

References: 178
Twelve (+Bonus) amazing YouTube Channels to Learn Python Programming for Free 179
    Clever Programmer 179
    Anaconda Inc. 180
    Talk Python 180
    Christian Thompson 180
    CodingEntrepreneurs 181
    Corey Schafer.. 181
    Chris Hawkes 181
    Enthought 182
    Real Python 182
    Sentdex (Harrison Kinsley)... 182
    Python Basics | Learn Python Programming 183
    Telusko 183
  Bonus .... **Error! Bookmark not defined.**
    Al Sweigart 183
    PythonBytes 183
A Simple Approach to Define Human and Artificial Intelligence................... 184
  A long time question… 185
  What is Intelligence? 185
  Types of intelligence.. 186
  Emotional intelligence 188
  Artificial intelligence 188
  Human Intelligence vs. Artificial Intelligence 189
  Conclusion 191
  References: 192
What is Predictive Analytics, and how can you use it today? 193
  What is predictive analytics?... 194
  How does predictive analytics work? 195
  Importance of predictive analysis for companies... 195
  About Predictive Analytics, Big Data, and Business Intelligence. 196
  What are predictive models?... 197
  How to Do Predictive Analytics in 7 Steps.. 198

Definition of objectives	198
Definition of analysis goals	198
Data collection	199
Data preparation	199
Data analysis	199
Modeling	200
Monitoring	200
Three examples of Predictive analysis software	200
Microsoft Power BI	200
Adobe Analytics	201
Tableau	201
Five practical examples of how predictive analytics can be applied today.	201
Customer Churn forecast	202
Campaign optimization	202
Lead segmentation	202
Customer Relationship Management (CRM)	202
Fraud detection	203
Risk management	203
Conclusion	203
Read more about it…	204
References	204
Will My Daughters Ever Drive a Car?	205
The dream of a new mobility	206
Safety as a horizon for new mobility.	206
Autonomous Vehicles will drive better than you do.	207
Rethinking our cities	208
What are the advantages of AVs?	209
And what about the driver's license?	209
Conclusion	210
Read more about it	211
Thoughts about the Need for More Diversity and Inclusion in AI	212
Why does the lack of diversity in Artificial Intelligence can be an issue?.	213
Artificial intelligence & diversity	214
Can AI itself help us to solve the lack of diversity in AI?	215

Conclusion	216
Read more about it	216
References:	217
What is Prescriptive Analytics, and what can it do for your business.	219
Why prescriptive analysis matters to your business?	220
Conclusion	222
Other articles you may want to read	223
Introduction to the Future with 5G, AI, and IoT	224
It is more than the internet…	224
Getting to know a little bit more about 5G	225
Getting to know a little bit more about AI	226
Getting to know a little bit more about IoT	227
What will be the relationship between AI, 5G, and IoT?	228
Conclusion	229
References	229
These are some of the best YouTube channels, where you can learn PowerBI and Data Analytics for free.	231
Why PowerBI?	232
Microsoft PowerBI Channel (243K subscribers)	232
PowerBI Tips (6.38K subscribers)	233
A guy in a Cube (150K subscribers)	233
BI Elite (28.4K subscribers)	233
Curbal (57K subscribers)	233
Enterprise DNA (41.6K subscribers)	234
RADACAD (13.2K subscribers)	234
Pavan Lalwani—POWER BI (7.23K subscribers)	234
Other resources	235
Microsoft guided learning	235
Microsoft PowerBI blog	235
Microsoft PowerBI Webinars	236
Official Power BI Community	236
Conclusion	236
One more thing…	237
What is TinyML, and why does it matter?	238

A small device for a tremendous impact. 238

Cracking the small ML 239

Where can you learn more about TinyML? 241

Conclusion 242

One more thing… 242

    References 242

The Best MIT Online Resources for You to Learn AI and Machine Learning for Free 244

Why should you learn AI and Machine Learning? 245

Are you ready? 245

    Artificial Intelligence 246

    Introduction to Machine Learning 246

    Machine Learning 247

    Introduction to Computational Thinking and Data Science 247

    Techniques in Artificial Intelligence (SMA 5504) 247

    Mathematics of Machine Learning 248

    Introduction to Deep Learning 248

    Machine Learning for Healthcare 248

    Deep Learning for Self Driving Cars 249

Conclusion 251

    Bonus .... **Error! Bookmark not defined.**

    Enter Julia, a rising start on AI development. 249

    Discovering a brave new planet 250

Read more about it… 251

References 252

Photo References 269

Afterword 270

About the Author 271

    Social Media Links 271

Copyright and Disclaimer 272

    Free Opinion Disclaimer 272

Photo References

1 - Cover - Blue vector created by stories - www.freepik.com
Figure 1-a- Allegorical Portrait of Dante - Agnolo Bronzino - Public Domain 6
2 - People vector created by pch.vector - www.freepik.com 7
3- People vector created by freepik - www.freepik.com 11
4 - Education vector created by vectorjuice - www.freepik.com 17
5 - Abstract vector created by pch.vector - www.freepik.com 24
6 - People vector created by pch.vector - www.freepik.com 34
7 - People vector created by stories - www.freepik.com 40
8 - People vector created by freepik - www.freepik.com 43
9 - Illustration vector created by pikisuperstar - www.freepik.com 48
10 - People vector created by pch.vector - www.freepik.com 51
11 - Business vector created by teravector - www.freepik.com 56
12 - Technology vector created by vectorjuice - www.freepik.com 59
13 - Technology vector created by vectorjuice - www.freepik.com 65
14 - Design vector created by freepik - www.freepik.com 70
15 - Illustration vector created by stories - www.freepik.com 74
16 Infographic vector created by rawpixel.com - www.freepik.com 79
17 - Technology vector created by vectorjuice - www.freepik.com 84
18- Car vector created by vectorjuice - www.freepik.com 89
19 - Car vector created by vectorjuice - www.freepik.com 96
20 - Car vector created by vectorjuice - www.freepik.comHow 108
21 - Background vector created by rawpixel.com - www.freepik.com 117
22 - Technology vector created by stories - www.freepik.com 130
23 - School vector created by vectorjuice - www.freepik.com 138
24 Social media vector created by stories - www.freepik.com 141
25 - Business vector created by pch.vector - www.freepik.com 152
26- Blue vector created by vectorjuice - www.freepik.com 156
27 - Infographic vector created by katemangostar - www.freepik.com 162
28- Car vector created by vectorjuice - www.freepik.com 171
29- Business vector created by katemangostar - www.freepik.com 175
30- Work vector created by stories - www.freepik.com 180
31 - Infographic vector created by katemangostar - www.freepik.com 185
32- Cartoon vector created by vectorjuice - www.freepik.com 194
33 - Car vector created by vectorjuice - www.freepik.com 207
34 - People vector created by pikisuperstar - www.freepik.com 214
35 - Business vector created by stories - www.freepik.com 221
36 - Business vector created by katemangostar - www.freepik.com 226
37 - Business vector created by stories - www.freepik.com 233
38- Cartoon vector created by vectorjuice - www.freepik.com 240
39 - Business vector created by vectorjuice - www.freepik.com 246

Afterword

Thanks for making it this far, and thanks for the trust.

I hope that much of the information and knowledge shared in this volume will be useful in your work, in your daily life, and your future.

I believe in the ability of technology, especially in Artificial Intelligence, to improve our lives and the lives of millions, perhaps billions of people in the world, if applied with equity and inclusion.

See you online.

Thanks

Jair Ribeiro

About the Author

Jair Ribeiro was born in Brazil and has lived in several countries since 1998; father of three daughters, he lives in Poland since 2016.

Combining Business Analysis, Data Science, and Artificial Intelligence with his multicultural background, ethics, and design, he is always exploring AI's role in a human-centered world.

With a broad leadership and multicultural background, Jair Ribeiro has been performing the role of AI Evangelist in several media (mainly on LinkedIn with more than 25.000 followers) and has been an active international speaker, presenting in several conferences and Summits, engaged in popularizing AI from the business point of view.

He is also very engaged in several activities related to Leadership, Multiculturalism, Volunteering, modern workplace, and International Career Management.

Social Media Links

Medium: https://jairribeiro.medium.com
LinkedIn: https://www.linkedin.com/in/jairribeiro
Twitter: https://twitter.com/Liberoliber
Facebook: https://www.facebook.com/liberoliber2010

Copyright and Disclaimer

All rights reserved, including the right to reproduce this book or portions thereof in any form whatsoever. For information, address the publisher at jair.ribeiro@outlook.it

Although the author and publisher have made every effort to ensure that the information in this book was correct at press time, the author and publisher do not assume and, as a result of this disclaim any liability to any party for any loss, damage, or disruption caused by errors or omissions, whether such errors or omissions result from negligence, accident, or any other cause.

Free Opinion Disclaimer

The views and opinions expressed in this eBook and my articles are those of the author. They do not necessarily reflect the official policy or position of his current employer. Any content provided by the author is of his opinion and is not intended to malign any religion, ethnic group, club, organization, company, individual, anyone, or anything.

273

www.ingramcontent.com/pod-product-compliance
Lightning Source LLC
Chambersburg PA
CBHW060825220526
45466CB00003B/979